A Treatise on the Australian Merino Sheep

by John Ryrie Graham

with an introduction by Jackson Chambers

Self Reliance Books

Get more historic titles on animal and stock breeding, gardening and old fashioned skills by visiting us at:

Introduction

I am pleased to present yet another practical title on breeding and raising livestock.

The work is in the Public Domain and is re-printed here in accordance with Federal Laws.

As with all reprinted books of this age that are intended to perfectly reproduce the original edition, considerable pains and effort had to be undertaken to correct fading and sometimes outright damage to existing proofs of this title. At times, this task is quite monumental, requiring an almost total "rebuilding" of some pages from digital proofs of multiple copies. Despite this, imperfections still sometimes exist in the final proof and may detract from the visual appearance of the text.

I hope you enjoy reading this book as much as I enjoyed making it available to readers again.

Jackson Chambers

TO

P. A. JENNINGS, Esq., M.P.,

OF

WARBRECCAN,

REPRESENTATIVE OF THE MURRAY DISTRICT IN THE

PARLIAMENT OF NEW SOUTH WALES,

AND

PRESIDENT OF THE PASTORAL ASSOCIATION

OF DENILIQUIN,

I DEDICATE THIS BOOK.

JOHN RYRIE GRAHAM.

TABLE OF CONTENTS.

———

TO THE READER.

IT were scarce seemly to meet, as I believe I am about to do, a courteous reader, without a courteous salutation, which, therefore, I here respectfully tender to all who honour me with a perusal. But thoroughly to relish and assimilate the contents of a work necessarily didactical, it appears to me essential to know what manner of man was he who wrote it. My autobiography for this purpose may be comprised in a dozen or two of words. Born some fifty and odd years ago, within "cooey" of a New South Wales sheepyard, I have, with few and brief intervals, ever since dwelt in the midst of Australian flocks. I am still so occupied, and so long as physically I continue robust, and mentally unimpaired, it is my intention to pursue the avocation.

In a work of this kind, resting its claims to notice solely upon the worth, whatever that may amount to, of my own personal experience, some passages must necessarily have an egotistical or dictatorial twang, and these, amongst others, will evoke dissent. If such dissent consist merely in the collision of opinion, without which truth is rarely elicited, one at least of my aims will be attained, and no person will more sincerely hail the subversion of my views than I shall myself, provided only that there be established in place of them, maxims more practical, sound, and useful. To be exempt from criticism I neither expect nor desire, for I am not too young to acknowledge an error, nor too old to be instructed; and the reader, generous as well as courteous, will not think the worse of me for the candid acknowledgment that, day by day, I can find much to learn, which, night by night, I strive to digest.

Very little, indeed, I owe to books for what knowledge of my subject I may possess ; yet I fully recognise the value of works such as those of Youatt, Spooner, Loudon, Henry, Stephens, and a few others. But a great book is often a great evil; therefore I have declined to dilate my small one with quotations.

In fine, I should never have dreamt of coming before the public in a way so apparently pretentious, and so certainly perilous, but for the kindly urgent suggestions of my patrons and friends, whose desires it is neither my inclination nor my interest to oppose.

JOHN RYRIE GRAHAM.

19 PARK STREET,
    EMERALD HILL.

A TREATISE

ON THE

AUSTRALIAN MERINO.

CHAPTER I.

THE CAMDEN, OR McARTHUR'S SHEEP.

Much has been said and written concerning these sheep, their excellences as compared with the more modern Australian Merino, their comparative value as wool-producers, and the part they have borne in forming or modifying the existing breed. These sheep and their patriotic breeder constitute a chapter in the history of Australia, important and interesting perhaps indeed to the whole world. From a worthless class of sheep, Mr. McArthur, with consummate skill and judgment, succeeded in producing sheep of the greatest value to commerce. True, he was aided by the finest wool-growing climate known, but without the acute perception, judgment, and skill which Mr. McArthur brought to the task, no merely natural advantages could have availed. When Mr. McArthur succeeded in obtaining his English Merinos from the royal flock of George III. there was not, perhaps, in the Australias another gentleman of social position so distinguished that he could have accomplished the task. At first he had very great difficulties to contend with, and was often advised to relinquish his pursuit; had he done so, who shall say what would have been the effect on the progress of civilisation of the world; but he persevered, and he had his richly-merited reward.

It was in the year 1826 that I first saw these celebrated sheep; they were much smaller than the Australian Merino of the present day, and each had a small leathern collar around its

B

neck. From the impression remaining on my mind, after the lapse of 44 years, I should think that the ewes I saw would not weigh more than from 30lb. to 34lb. each. Although small, they were extremely compact, low set, short-legged, bodies close to the ground, and as much alike as the grains in a handful of duck shot. Their wool was exceedingly fine, but certainly not dense. The country they were feeding upon was inferior, and heavily timbered with gum, peppermint, etc., excepting a few small paddocks which had been cleared ; and, as might be expected, the rams when sold and placed upon other runs, having better and richer food, always begat lambs, in point of size, far superior to themselves. At the time I refer to, and for many years after, these sheep were undoubtedly the best in the Australian colonies, and could Mr. McArthur have produced three times the number of rams he did produce, they would have met with a ready sale. The only objection made to them was their diminutive size, few breeders at that period having had sufficient experience to observe that many of them were defective on the back and across the loins. Not for a moment do I pretend to infer that I discovered or even dreamt, at the time, of this being a defect. Like many young Australians of that day, to believe in these sheep was a part of my creed, and I certainly entertained no expectation of seeing them equalled, much less excelled. This increased size in the progeny of the Camden rams was due, doubtless, to what that eminent Victorian breeder, Mr. Robert McDougall, would designate an increase of natural comforts. When Mr. Riley, of Raby, about the year 1828 or 9, imported the Raby sheep, and commenced breeding, he would occasionally point out to his visitors the necessity of density as well as fineness, and grasping the fleece of a thin-woolled sheep across the loins, which would scarcely half fill his hand, he would then select a sheep with a fleece so dense that the hand would not close upon the quantity grasped. It was then I began to perceive the defect of the Camden sheep in this particular, but fineness was the great desideratum, and the Camden were undoubtedly the finer fleeced. In fact, the object aimed at by the proprietor of the Camden sheep was fineness, all other considerations being secondary only, whereas the Raby sheep, although still very fine, were bred with a view to density and length of staple. I know many squatters and persons connected with the squatting interest, when the Camden sheep were in the zenith of their popularity, who now assert that we never had, nor ever shall have, sheep to equal them ; but, whilst gladly conceding great merit to the Camdens, I fearlessly assert that we have on many stations in Victoria sheep infinitely superior to any ever bred on Camden. As regards fineness of staple, we are fully equal to them—I will not say superior. In point of carcase, length of staple, weights of individual fleece, and density, we leave them far behind. The rams of Mr. John Aitken, the Messrs. Learmonth, Mr. Currie, and not a few others, would cast the best sheep ever bred by Mr. McArthur into the shade. Indeed, I consider that I should pay Mr. Learmonth, for instance, but a left-handed compliment, were I to declare that I thought his sheep equal only to Camdens. On the contrary, I submit that Mr. Learmonth's sheep ought to be superior. Consider the sorry lot Mr. McArthur had to commence with, and the long years it must have taken to eliminate the combination of bad qualities possessed by these pristine animals; whilst on the other hand, Mr. Learmonth and his compeers have skilfully availed themselves of the opportunity of using sheep, both ewes and rams, improved to a degree which it had cost Mr. McArthur many years of toil to attain. In a word, our ram-breeders commenced just about where Mr. McArthur left off. Mr. Learmonth, doubtless, commenced with the best sheep that could be purchased, and I believe he has been carefully breeding rams for nearly 30 years; surely, during that extended period, a gentleman of Mr. Learmonth's admitted ability, wide experience, and enlightened acquaintance with the true principles of breeding, could not fail to produce a highly improved race of animals. It is, therefore, I submit, perfectly absurd to compare the Camden sheep with those to be found in possession of many of the more distinguished breeders of Victoria at the present time.

NOTE.—The substance of the foregoing article appeared originally in the columns of the *Australasian.*

CHAPTER II.

ON CROSSING THE AUSTRALIAN MERINO, AND ITS EFFECT ON OUR WOOLS.

THE crossing of the Australian Merino, with the coarse English breeds, Leicester, Cotswold, Teeswater, South Downs, and Oxford Downs, also with the French Rambouillet, and the German sheep, has had a very disastrous effect upon our wools. The subject is one of more than common interest and moment. I shall therefore endeavour to trace it from its origin to the present day. In the year 1797, Captain Kent brought from the Cape a small number of pure Merino sheep, which were distributed between himself, the Rev. S. Marsden, of Paramatta, Mr. Cox, (father of the Mudgee Coxes,) and Captain McArthur, five ewes and one ram falling to the latter gentleman's share. What became of Captain Kent's lot I cannot ascertain; Mr. Marsden's sheep were sent to his farm, near Liverpool, and their progeny, I believe, having ultimately become the property of Mr. Betts, were many years subsequently sent up to the Bathurst district. Mr. Marsden was a good farmer, and noted for a peculiar breed of very superior dairy cattle; but he was never eminent as the breeder or possessor of fine woolled sheep, so that I conclude he was unsuccessful with his imported Merinos. Fortunately for the colony, Mr. McArthur perceived the value of those which had fallen to his share, and with these sheep, and a few English Merinos also imported, he laid the foundation of the Australian Merino. Of the fate of Mr. Cox's sheep I have not had means of positively speaking, but judging from the fact that all the members of that family, especially Mr. Edward Cox, have ever since been noted for the possession of superior flocks, I am inclined to think that the blood of Captain Kent's importation is still to be found at Mudgee. From 1803 up to 1828 or 9, we hear little of imported sheep; and meanwhile, under the intelligent management of Mr. McArthur, it took some twenty-three years to perfect the pure breed of Australian Merinos. In 1828 or 9, Mr. Riley of Raby Park, Mr. Walker of Wallarawang, and Mr. Bettington of Bathurst, imported some Saxon Merinos. In 1826, the Australian Agricultural Company, having obtained a grant of one million acres, at Port Stephens, Newcastle, and on the Peel River, commenced sheep farming on a large scale, first under the management of Mr. Dawson, and afterwards of Captain Parry (who, by the way, was styled "Commissioner," and received a salary of £5,000 per annum, when our Governor, Sir Ralph Darling, was getting £3,000 only). In 1834, the Company imported Saxon Merinos, and gave a marked impulse to squatting pursuits, by the high prices they paid for stock of all kinds to commence upon; £12, for instance, was given for ordinary breeding cows; £5, and even £6, per head for ewes. This, together with the higher appreciation, and therefore steadily augmenting value of our wools in the home markets, induced the above gentlemen to import rams freely, with the view of still further improvement. Riley and Walker's importation was, I think, a joint speculation; but however that may be, the imported rams were certainly the best I ever saw, of which, when he came to dispose of their progeny, Mr. Riley had some proof, in the high prices they brought, and the great estimation in which they were held. They had no leather necks, nor wrinkles on shoulders or hind-quarters, in fact their skins fitted them and they filled their skins. I know of no sheep that they resemble so much, in point of form, as some of the best bred Australian Merinos of the present day. They had not perhaps the exceptionally perfect symmetry of Old Grimes, but they displayed exactly the form of Mr. Learmonth's, Mr. Shaw's, or Mr. Currie's Australian Merinos. Their fleece could not be surpassed.

Mr. Riley bred from McArthur's ewes; and many of the Bathurst and Mudgee breeders are indebted to that stock for the present superiority of their flocks. From about 1829 to 1840 the Australian wools had a character so uniform and fixed that an English wool-broker or sorter could with certainty select by the touch alone from a bale of others, a Botany Bay Fleece, as they were then termed. The best wool in the world, however, was not good enough to content the squatters of that day, and in the attempt to put a Merino fleece upon a Leicester carcase they nearly ruined the colonies, and not a few of themselves entirely.

If sheep-owners desirous of increasing the size of their Australian Merinos will carry out the system initiated by Mr. William Lee, senior,* of Bathurst, they will doubtless arrive at the successs that gentleman attained. Mr. Lee, I may state, is one of the oldest as he is one of the most successful and practical squatters in New South Wales, he having followed the business of sheep-farming for more than sixty-years; and as the experience of such a patriarch is

* Mr. Lee is widely known as the breeder of the celebrated race-horse "The Barb."

worth all the precept that could be written, I will briefly narrate the history of Mr. Lee's operations. At Larry's Lake, near Wellington, Mr. Lee had a run carrying say about twelve thousand sheep, and each succeeding year he invariably disposed of his surplus stock, whatever prices might be ruling. Say that out of these twelve thousand sheep he had five thousand lambing ewes, whose season's increase amounted to four thousand five hundred of equal sexes, this would be the total number of sheep he would have to dispose of, in order to keep the stock on his run at or near the normal number of twelve thousand. Half of the increase being ewe lambs, he would possibly reject a moiety of them, or 1125; he would also cull from the ewes say one-half, or 2250, and the balance of 1125 he would take from his wether flocks, thus completing the number of 4500. He, of course, selected the largest and best woolled sheep to breed from every year, and by steadily pursuing this simple and rational system for a number of years, he ultimately bred the largest framed Australian Merinos I ever saw. Mr. Charles Smith, the eminent butcher of Sydney, once told me that he purchased a flock of ewes from Mr. Lee, which averaged over 70lbs. weight, and the wool of these ewes was sold in Sydney, the year preceding, for 2s. 9d. per lb. Mr. Lee invariably kept his run what would now be considered understocked; his sheep never knew a check from the day they were lambed until the day of their death or sale. I have often heard him say. "My sheep are lambed fat, and I take care to keep them so; if lambs are not worth feeding, they are not worth rearing."—(*Vide* Loudon's "Encyclopædia," Sec. 6482.) It appears then that the smallness of our sheep can be easily obviated by good feeding, attention to their "natural comforts," and a proper selection for breeding. Feed them well, and they will pay for it; half-feed them, and they will not half-pay you; neglect them, and they will ultimately ruin you.

There is no doubt that a large-sized Merino consumes more food than a small one, but the small ones for many parts of our thinly-grassed countries, are certainly best adapted, as they are more active and can fill themselves in less time, consequently they can keep themselves in better condition in times of scarcity. It just comes to this, that in poor thinly-grassed countries you must not try to produce sheep larger than your country will keep and fatten; in richly grassed countries you may and can breed them to almost any size you like. Still I am of opinion that we do not, in this country, require a large, heavy sheep; but rather one that is small and compact, short in the leg, deep in the body, yet sufficiently active to walk after his food without distressing himself, when, as is too often the case, food is not over-abundant.

Up to 1839, I do not recollect any importation of the Leicester or other English breeds, and I consider that about this time, and mainly for this reason, the Australian Merino attained its highest point of excellence. Time, climate, and intelligent management had all been employed in eradicating the legion of bad qualities which characterised the original stock, and the whole of the sheep in the colony were types of one class, differing only from the operation of local influences, and the degree of intelligent care bestowed upon them by their various owners. Of course some flocks were much better than others, but all had the stamp of the Australian Merino impressed upon them; there were no mongrels among them, and although nearly every squatter had been breeding for numbers only, irrespective of quality, yet such was the surprising durability and vigour of the blood that neither cupidity, neglect, or total ignorance of the true principles of breeding could destroy the character of the wool. From about 1835 to 1840, a perfect mania for sheep-farming had taken possession of military officers, gentlemen of the mercantile navy, lawyers, clergymen, merchants, and others who had capital to invest, or credit to float a set of bills. Sheep and their wool were the all-absorbing topic of the period, and immense numbers of sheep and stations became the property of gentlemen who had not the slightest idea of stations, sheep, or their management. It is, therefore, not at all surprising that, from this time, the ruin of the Australian Merino commenced. All so far was doing well, but the would-be sheep-breeders thought they could do better, if they could only succeed in placing the Merino fleece upon the Leicester carcase, (they might as well have tried to grow it on a bullock,) and with this view, Messrs. Icely and Rodd, of Goombing, near Bathurst, imported, I believe, in 1840, the first Leicester sheep. The arrival of these animals, ewes and rams, created quite a sensation in the district; every one was eager to see them, myself among the number, and when at last I found the necessary leisure and opportunity I formed an opinion of them which I have not since changed, viz., that they were very fine animals, but it was a pity they had not left them "at home." The importers, however, commenced breeding from them by crossing with some of their best Australian Merino ewes. The first cross did certainly gain something in size, and looked well until they became two years old, but the increase of bulk was palpably at the expense of the wool, which had become harsh and loose, and on the back had a "mushey" feel and appearance,

Notwithstanding this, there was quite a rush for rams, the progeny of these sheep, and Mr. Rodd succeeded in disposing of them at extreme rates, as fast as he could breed them. But when, after lambing, these mongrel ewes commenced to throw off their wool from their bellies and points, retaining in fact no wool except a little thin stuff on the ribs and back, one would have thought the eyes of the sheep-owners would have been opened to the error they had committed; but no such thing; they had obtained size, and that was enough for them. Every year, as the sheep grew older they became worse, until at the expiration of about seven years it was discovered that these cross-breds inherited neither the fattening properties of the Leicester nor the wool-bearing qualities of the Merino; in a word, they were pure in one respect only,—they were pure mongrels. Meanwhile, before this sad truth was discovered, three-fourths of the sheep in the district were completely ruined. About this time Mr. Denne, of New England, imported some Oxford Downs sheep, which led to about similar results. About 1844, the Australian Agricultural Company imported some either Leicester or Lincoln sheep, but the gentleman in charge of the Company's sheep, with an amount of decision and judgment that does him infinite credit, declined to breed from them, and disposed of them for what they would fetch. Yet in spite of all these total failures, and the acknowledged injury the introduction of these animals had inflicted upon the great staple of the colony, her wools, many perverse and misguided individuals continued largely to import them, but so far fortunately for the country, many of the worthless brutes died after the first or second year of their colonization. I have heard fine-woolled sheep objected to—even in Australia—on the score of their delicacy. (*Vide* Youatt, p. 179, ed. 1864.) Compare the facts stated in other parts of this work with that just mentioned of the early decease of the cross-breds, and then let the enlightened reader judge between the two. As a specific exemplification of the hardihood of the Australian Merino, I would mention that in 1839, a large number of them were travelled over the Snowy Mountains, lambing as they went, and more than once compelled to remain some days stationary on account of the snow, yet they reached their destination, South Australia, with a loss of only 5 or 6 per cent. of the ewes, and a good number of lambs alive and thriving. Fine-woolled sheep are sometimes objected to because it is thought that a fine fleece must, as a matter of course, weigh less than a coarse one; this is an error, a fine-woolled fleece, if dense and long, will weigh more than a coarse one, of course both being Merino. It is not to be supposed that a fine-woolled Merino fleece, be it ever so good, will weigh as much as a Cotswold, Leicester, or Lincoln.

It is certainly singular that after the Colonies have had no less than thirty years experience of the dire results of crossing with these sheep, people are still to be found who advocate the continuance of the practice. Even in point of size it can do us no good, which, within the limits set to us by climate, pasturage, etc., we cannot better effect without extraneous aid. If our Australian Merinos be too small for us (I contend they are not), we can easily get them larger; feeding, an increase of "natural comforts," and a persistent selection of our largest ewes and rams, to breed from, will do this for us; but, for reasons which I have elsewhere assigned, I doubt very much whether an increase of mere bulk would be an improvement. At all events, if such increase be desirable, it must not be obtained at the cost of the wool; in other words, it must not be secured by crossing with large, coarse-woolled sheep, whether for carcase or for wool, the difference between our Merino and any coarse-woolled sheep whatever is too great. The Saxon Merinos imported by Messrs. Riley, Walker, Bettington, and the Australian Agricultural Company, were not of the trashy description that has of late years been too freely introduced into Melbourne; the former were not got up for the occasion, with two or three years wool left upon them, and as much grease as wool, but were purchased by gentlemen who were thorough judges, who went to Germany on purpose, and there selected and paid a good price for the sheep that suited them. To the latter, there were and are in Victoria many sheep equal, if not superior, as regards either wool or carcase; facts which I shall be compelled to reiterate in several parts of this volume. Let me be clearly understood; suppose I had the choice of two rams, one an imported Saxon Merino, the other an acclimatised Australian Merino, both being alike as regards wool, form, age, etc.; I should certainly take the Australian Merino, because he had arrived at his state of perfection in the very country I wanted him for, and I should reject the German because he had acquired his perfection under circumstances and conditions which could be no guarantee of his fitness for effective service in Australia; the former having acquired his form, condition, wool, etc., unassisted by anything but our natural climate and grasses, whereas the German from his birth had been housed, tended, fed, and cared for like a delicate baby. The tenderness and delicacy thus conferred upon him, he would, for some time at least, transmit to his offspring, whilst in the Australian Merino you have the very animal you require, in the utmost natural perfection. I do not mean to infer that the German sheep, or at least his descendants, may not be made quite equal to the Australian, but what man in his senses would spend seven, ten, or more years in the endeavour to effect an object which could be as well or better carried out in a few days? Why buy imported German rams, which can only in the course of many years produce sheep *perhaps* as good as those which Messrs. Cummings, Russell, Simpson, Learmonth, Currie, Shaw, and several others will supply to you next week and next door? But besides this it is not every German sheep that is susceptible of acclimatisation; numbers of them will degenerate in place of improving; this may be easily verified by comparing the amount of wool shorn from them in the first with that obtained in the second year. In nearly all of them that I have had relations with, I have detected a gradual diminution in weight of fleece every year, and the few that gained in weight of wool were clearly exceptions only. I am speaking now of the Germans that have been imported within the last twelve or fourteen years; and not of such sheep as were imported by Messrs. Riley, Walker, and Bettington; these sheep were good when they came to Australia, and remained good, but still they were never so good as some of their increase.

A further and not a light consideration is the well-known fact that no German breeder of eminence ever, under any consideration whatever, will part with the best sheep of his flocks; so that, after all, the Germans we do get are but second, perhaps, considering the demand by other European breeders, only third rate. It is quite true that the best of our present flocks owe, at least, the origin of their excellence to the Saxon Merino, but without time, a climate surpassingly favourable to the growth of wool, skilful management, and a profound knowledge of the principles of breeding, the result would have been far different indeed. Perhaps, I would scarcely have ventured to express, on this subject, opinions so decided, but for the certainty I possess that similar convictions are those of the more experienced and enlightened among our flock-masters; who, however, refrain from putting them prominently forward, lest they be falsely charged with unworthy motives. After the expenditure of enormous sums which have been applied to the purchase of German and French sheep, during the last fifteen years, I would respectfully ask to be shown a single station where, in consequence of their introduction, any improvement, either in quality of wool or weight of fleece, can be proved to have taken place. Of course, it is not every flock-master who possesses the knowledge and experience which are absolutely essential for the successful selection of sheep for breeding purposes. It demands time, tact, patience, skill, experience, and no small amount of cash. Ram-breeders, to meet the times, have been selling large numbers of very inferior animals, at prices as low as from twenty to thirty shillings each. Now, a ram whose intrinsic value is only thirty shillings, is not worth breeding from ; no one can afford to breed really good rams at that figure, therefore the transaction is a mistake on the part of both buyer and seller. It would be much more economical, on the part of the buyer, to pay £10 for a really good ram, and it would be much better for the breeder if he declined to sell any ram that he did not consider worth, at least, that sum. Any ram not worth that limit should be castrated. The price of McArthur's rams ranged from £15 to £50, Mr. Riley's from £15 up to £32, and they were well worth the money ; but both these gentlemen, of course, had rams selected for their own stud-flocks, which no amount of money would have purchased. In my opinion, the best rams imported to Melbourne, of late years, were some American rams, the property of Mr. Campbell. In point of symmetry they were excellent, but the wool of many of them was wiry and harsh to the touch. A few were really excellent sheep, and all were superior to the trashy Germans we have lately seen. Two rams purchased by the Messrs. Peppin, of Morago, on the Edward River, and one by Mr. Younghusband, were, in my opinion, about the best of them.

The Leicester or other coarse breeds of British sheep (Youatt, pp. 320, 325), may perhaps pay in certain localities ; as on the rich, succulent pastures of Warrnambool, Camperdown, Colac, the Little River, or Werribee ; where the paddocks are laid down with English clover, lucerne, etc., and whence the proprietors are in a position to send large fat lambs early into market. It is said that Mr. Clarke did very well with his Leicesters on the natural herbage of this country, but it must be borne in mind that Mr. Clarke had a large tract of land peculiarly adapted for them, namely, Dowling Forest, near Ballarat, which was always kept considerably understocked, for the purpose of fattening those sheep. They are, however, too large to do well, as a rule, upon our thin indigenous grasses ; they are not sufficiently active to walk after their food, and when crossed with our Merino they lose the propensity to fatten early. In my opinion they should be kept pure to pay well, even under favourable circumstances, in Australia.

As regards the wool of the cross-breeds, it is neither one thing nor the other ; it retains the good qualities of neither Leicester nor Merino ; it is light, tender, and destitute of character. In effect, I am of opinion that the introduction of the Leicester sheep has done more injury to the entire wool-growing interests of the colony than scab, catarrh, foot-rot, or all the other ills which sheep are heirs to—at least in Australia.

The Rambouillet, too, has done us no good, because, without artificial sustenance, it is too large for the country; neither do I think it disposed to fatten early, even on the best of feed. The wool, certainly, is not nearly so objectionable as that of the Leicester, although not equal to that of our acclimatised Merino.

The introduction of the trashy German sheep, after our discouraging experiences of the past fifteen years, will soon cease, and I should think that few more will ever be sent to this country, unless by the most forlorn or reckless of speculators.

Of all the imported sheep, those of our first-cousins, the Americans, are the best; nevertheless, their Australian purchasers paid too dear for their whistle, simply because, for much less money than was paid for the Americans, they might have bought, almost at their very doors, far better acclimatised sheep.

The restricted size of the Australian Merinos is the principal objection urged against them; they are, it is said, too small; but it is my firm conviction that in this very smallness consists a great part of their national value. I mean, of course, their relative value as compared with the English Leicester. That the Australian Merino has, during the past twenty-five years, degenerated in point of size, I admit; neither is it at all surprising, when we take into consideration the scanty feeding they have of late years been accustomed to. Many flock-masters are beginning to recognise the advantages of light stocking, taught by bitter experience of the converse; and if light-stocking be persevered in, we shall soon have our Australian Merino wethers, weighing once more their 80 or 90lbs., and both the quantity and quality of our wool restored to their pristine superiority.

But this can only be effected by breeders who possess land capable of carrying the larger sheep, and bringing it to the highest perfection. And here I must reiterate the fundamental axiom, viz., that the desired increase of size must on no account be sought by means of admixture with a larger breed of sheep; purity of race will be found, for this purpose, no less essential than it is for the production of the best and most remunerative wools. In the remarks on "Culling Ewes and Rams," my readers will find a conclusive example of what may be effected by preserving purity of race, as practised most successfully by Mr. Lee of Bathurst.

I am rejoiced to observe that at length a reaction is taking place with respect to this rage for importing coarse-woolled sheep and Germans. If we can only succeed in boiling-down, preserving for exportation, or otherwise disposing of all sheep now in our flocks bearing the Leicester impress, and can close our ports to the introduction of any more German trash, we shall be in a fair way to re-establish our former reputation as the producers of the most valuable wool in the known world. I have dwelt especially on the dangers of over-stocking, and on injudicious crossing, or in any way mixing the blood of the Australian Merino with any other description of sheep, whether Leicester, Negretti, or Rambouillet; but I will add, that if all I have written be followed to the very letter, and these two most important points neglected, the sheep-owner may rest assured that he will never possess anything but a lot of half-starved, ill-formed, hybrid mongrels; that is, if the sheep be fed only on our natural grasses.

CHAPTER III.

ON BREEDING AND INTERBREEDING.

THE controversy which has, one may say, for centuries raged between the advocates of "in-and-in breeding" and the opponents of that system, would appear to be now in a fair way to cease; and this most desirable consummation has been so far achieved simply by the fusion of the two parties to the squabble. On the one hand, "*no* interbreeding" was the war-cry; on the other, "*any* interbreeding;" but at length some angel of peace suggested the adoption of the practice of careful "selection," under any circumstances, whether of consanguinity or otherwise, and now the in-and-in breeders admit the great benefit to be secured by an occasional judicious cross, whilst the other party will not object to the coupling of near relations, so long as there be no congenital or hereditary defect or disease on either side.

For my own part, I am well pleased at this, for I have ever considered in-and-in breeding, accompanied unswervingly by judicious "selection," to be the true secret of successful sheep-farming; but unless great care and skill are unremittingly applied to its every detail, the sheep-owner may work for his lifetime without success. If he have good sheep to commence with, a single false step may ruin them; if his sheep be inferior, the slightest error will make them worse. Let us suppose we are called upon to deal with an average lot of sheep; just an improved lot, which have not been classed, but are supposed to be as good as their neighbours. Well, we want to make them better. As a first step, I would carefully class them into first, second, and third quality. From the first, I would breed; from the second I would also breed, provided the station were understocked, but I would take great care that a distinguishing ear-mark was put upon them and their lambs. The third lot should never be bred from, but they should be fattened on the station, or sold as stores for any price they would bring. The best of the rams (if there be any really good ones) should now be selected and put into the No. 1 ewes. If there be no good and suitable rams on the station, they should be purchased from some well-known breeder, and a fair price paid for them, for it is false economy to purchase inferior rams, at any price. Although we keep the No. 2 ewes, and breed from them, it will not be with the view of keeping them or their increase any longer than will suffice to stock the station with better sheep from our No. 1 flock. In fact, if the station were moderately stocked without them, it would be better not to breed from the No. 2 flock at all, but rather to fatten them off, or put the rams into them, and sell them in lamb. This latter course would not be difficult, because they would be all young, or at least sound-mouthed sheep, as all the old and broken-mouthed ones would be in Class No. 3. Say you have 5000 number ones, they would require 100 rams to serve them; and let me urge once more, purchase them from a reputable breeder, and pay liberally for them. Buying rams is clearly a reproductive investment, the more you pay for them the better they will (or, at least, should) be, and of course the more you will get for your wool. When the first dropping of these ewes, by the rams you have purchased, are about 20 months old, select 50 of the very best of them, and purchasing from the same party who supplied you with your first rams, a very superior one, put him to your 50 selected ewes, or to 100, if you can find that number of sufficiently good quality. If you breed from the 50 only, you will possibly have, at weaning time, 20 or 22 ram lambs; from these select the very best one, and when the proper time arrives, put him and his sire to the first selected 50, together with the 22 or so of his own age, provided the whole of the 22 are good; but should any of the 22 be inferior, in consequence of their dams throwing back, cull them at once, and keep only such as are faultless, or nearly so. By these means, as the old rams give up, you replace them with better stock, as the father of the young rams was worth (say) £50, whilst the old ones cost probably no more than (say) £5 or £6 each. Thus, in the first instance, you buy better sheep than your own; consequently, the first lambs of that drop must be better than their mothers; and these again, being put to a ram very much superior to either father or mother, must improve your flocks rapidly. And here I must put my readers upon their guard against the commission of an error which is even yet too common. You are possessed of a breed of sheep which you think rather highly of, in all respects save one, which we will call a deficiency in " closeness " of fleece. Either of your own notion, or, more probably, at the suggestion of some adviser, you post off to the station of Mr. Soandso, and from that gentleman purchase rams, solely because Mr. Soandso's wool has a reputation for that special quality—closeness—

wanting in your own flocks. But no special quality can be thus secured, as it were by one bound; and although, on examining the produce of the first cross, you may find an increase of closeness, the chances are that you will also find a whole colony of defects from which your sheep were previously exempt. At the same time, the generic character of your stock will be entirely changed by the admission of foreign blood; and after you have spent years and large sums of money in studying the special idiosyncracies of your own sheep, you will find that you must begin again to acquire the knowledge necessary for the successful management of the new race. Had you, imitating the judgment and patience of a McArthur or a Bakewell, selected from your own flocks only, the closest woolled sheep to breed from, you would have perpetuated all the good qualities already possessed by them, and, in a very few generations, added the only one in which they were deficient. (Youatt, p. 124). If rams you must buy, buy them only from some flock that has been long established, and if you do not afterwards breed your own rams, continue to purchase from the same breeder, so long as you own a sheep, or the flocks remain in existence. There has been of late a system adopted in Victoria, and in parts of New South Wales, which has already been productive of much evil, and of which, it is to be feared, the full effect has yet to be made manifest. A sheep-owner, and most commonly one quite recently become so, would purchase a few Negrettis from Germany, a few Rambouillets from France, and a few Americans from one of the United States, talk "jumbuck" at Scott's or the Melbourne Club for a few weeks, appear a few times on the wool-rialtos— Goldsborough's or Clough's—engage a manager with a spotless reputation for book-keeping in the office where his life had been passed; take a run up to the station, and set the said manager at work to breed rams for sale, with his heterogeneous sires, out of a more heterogeneous lot of ewes! On no consideration would I purchase rams from a flock only recently established, and more especially where three or four distinct breeds are kept on the station. Except in a very few peculiar localities, the Australian acclimatised Merino is the only sheep for Australia; and attempts, by wide crossing, to confer qualities upon him which our soil and climate forbid the sheep to possess, will continue to end, as hitherto they have invariably done, in miserable failure, retrogression, and loss. It is, we may rest assured, not alone possible, but easy, by means of continued selection, to produce, from the one flock, sheep of two or even three distinct types, bearing no resemblance to each other. It has been said that it is highly injudicious to breed from near relations having any tendency to hereditary disease. I should think so, indeed. But who, in the name of common sense, would dream of breeding from ovine " near relations," or no relations, either of whom had any tendency to hereditary disease, or glaring physical imperfection? For instance, no amount of perfection as regards the fleece could compensate for a delicate constitution, or for an imperfect physical organization; either of which is invariably transmitted to the offspring.

Against in-breeding it has been urged that it has a tendency to weaken the constitution; I can only say, that in the whole of my personal experience, and in the practice of others whose operations I have been intimately acquainted with, I never once knew this result to take place. By in-breeding, with careful selection, that eminent breeder, Bakewell, produced a race of sheep unequalled for the properties he desired, and by a close adherence to this system those properties have been perpetuated for upwards of fifty years. (Youatt on Sheep, ed. 1864, p. 315). At the commencement of this chapter, I endeavoured to point out the best mode of breeding rams from your own flocks; but the practice would answer well only under the most skilful and diligent management. If sheep-owners generally would cease breeding their own rams, —Youatt, p. 315—some peculiarly qualified persons would be induced to apply themselves exclusively to the practice of breeding rams for sale; and there is no doubt that this "division of labour" would eventuate in the production of the best possible article, at the lowest possible cost to the consumer. Whilst none but thoroughly competent persons would undertake ram-breeding solely, the squatter would be spared the trouble, expense, and risk of keeping and managing the stud-flock. In Riverina, at the present time, it is most difficult to obtain really good rams, unless by the tedious and expensive means of sending them round by sea from Melbourne to Sydney, and submitting them there to a quarantine detention of six long months duration. Now, rams can be bred in certain parts of the Riverina country quite as good as those bred anywhere else. Not, perhaps, in the very warmest parts of the district; but say in the neighbourhood of Corowa, on the Murray, or between Wagga Wagga and Gundegai, on the Murrumbidgee. In these localities the average temperature is much lower than in the more central portions of the district, say around Deniliquin, Wangonella, or Hay; and sheep could be bred in any number, having abundance of yolk; which, as

C

I have elsewhere shown, is of the greatest importance to Riverina sheep.

For a period of twenty-five years, the writer was engaged in testing the value of in-and-in breeding. Much controversy has taken place on the subject, and not long since, in the columns of the *Economist*, the discussion was renewed. Opinions diametrically opposed were then expressed, but those enunciated by Mr. McDougall and Mr. McKnight, my experience compels me fully to endorse. I may say that I never saw an entire flock of really good sheep that was not wholly composed of in-bred animals; and I scarcely think it possible to breed good sheep without having recourse to in-breeding. By in-breeding I do not mean indiscriminate breeding without selection; on the contrary, I mean breeding with judicious selection, that is, rejecting rigorously the faulty sheep, male and female, and breeding only from the perfect. After some years of practical experiment, I became so impressed with the paramount value of in-breeding, that I have on several occasions put the son to the mother, and produced a far better sheep than the father of the son was. Let me on this vital point make myself understood. I put together the best ram and ewe to be found on the establishment, and if I found the offspring of this couple was a ram lamb, and that he possessed the qualities I required in a more eminent degree than his sire, I would put him to his own mother, and rest assured that he would beget a better sheep than even himself or his sire; in fact, I have invariably found that the closer you breed, the more certain you are of the offspring resembling the parents.

Against the practice of in-breeding an argument is sometimes used, which must be noticed, as it not unfrequently imposes upon inexperience, namely, the inferiority of wild cattle and horses to those in a domesticated state. But the propounders of this argument entirely overlook the fact that *wild cattle and horses do not select,* and that the veriest wretch amongst them has every opportunity of begetting an animal like himself; whilst the original stock were of the most heterogeneous description, and every mongrel bull or entire horse which escapes and is not worth looking after, forthwith adds his quota to the mass of deformity, vice, and uselessness. I would not absolutely prohibit crossing, but I would never cross with any sheep but such as I knew had been in-bred. For instance, say Mr. Learmonth's ewes with Mr. Shaw's rams; or Mr. Currie's rams with Mr. Learmonth's ewes. In these cases both sires and dams are in-breds, and you might rely, if both of them suit your purpose, that they would throw lambs like themselves; they could not throw back, simply because they had been too long in-bred to permit of it.

I have often heard it remarked by persons engaged in the operation of drafting ewes for a stud-flock, "Oh, these ewes are a little faulty, but the rams are very good, and will counteract that." But if the rams are long in the staple, the ewes short—below the average—we surely cannot expect the progeny to ignore altogether the influence of the female parent, and bear wool as long as that of the male. An ewe should not be bred from if she be faulty; and on no consideration whatever should she be admitted into the stud-flock. I have always found it much easier to obtain good ewes than good rams, and for the stud-flock both should be perfect, a consummation which can never be attained without a long course of in-breeding from perfect parents. It will be observed that the two conditions upon which I principally insist are judicious selection and in-breeding, and without these in combination a squatter may work for a life-time unsuccessfully. Much, however, will still depend upon the knowledge possessed by superintendents and overseers. No doubt the great majority of these gentlemen are careful, industrious, and energetic, and, so far as the general routine of station management is concerned, they may be everything that could be desired; but there is not one in a hundred of those with whom I have had relation, who knows more about selecting sheep for breeding purposes than their grandmothers did.

The ordinary practice is to select for size. "If," say they, "a sheep be big enough he is good enough." I once had occasion to visit a station at the time when the superintendent was selecting rams for the ewe flocks—not a bright lot certainly; but observing one singularly objectionable ram, which I thought was a cross between a Rambouillet and a Cotswold, with as much hair on his throat and head as would make a judge's wig, I ventured to point him out, and ask the super. if he really meant to put him to the ewes? "Most assuredly I do," he replied, "see the size of him!" The animal was as tall, and in shape not unlike, a great half-starved kangaroo dog. Yet this gentleman has been a sheep-station manager for upwards of twenty years, and had entirely under his own control from 80,000 to 100,000 sheep.

Upon another large station I was asked to examine some rams, and gave a written report upon them; the owner informing me that it took about 207 fleeces to fill a bale of 250lb.! He thought (and so did I) that there must be something wrong, as he had spared no cost in purchasing rams nearly every year.   Upon proceeding to the station and conversing with the super., that gentleman coolly informed me that the difference between a good woolled sheep and a bad one was a "mere matter of opinion;" that he had taken every care to improve his sheep "by not breeding too closely;" that he never put the same ram twice to the same ewes, but exchanged rams with his neighbours, so as to avoid too close breeding.   This literal fact speaks for itself, and I will only add here the written opinion: "If he persist in his 'improved' method of sheep-breeding, he will soon manage to put his whole clip in one bale, and congratulate himself on the great saving in woolpacks."

In speaking of the various kinds of country used as sheep-runs, I have arranged them under four heads.   First, the eastern side of the main range, where gut-rot or coast-disease prevails ; secondly, the table-land and the western flank of the main range, where are found flukey grasses in abundance ; thirdly, the intermediate country, situate between the flukey and the hot salt-bush tracts ; and, lastly, the hot salt-bush country itself.   The intermediate is assuredly the preferable country to breed rams upon, provided the special locality chosen for the purpose be not in contiguity to the main range or its prinicipal spurs, where the flocks would meet with flukey grasses.   Freedom from disease, as well as good bulk of carcase, might thus be confidently reckoned upon ; whilst the temperature would be sufficiently cold to produce a good supply of yolk for the nourishment of the wool.   Mudgee is situate on the intermediate country, and it is often said that there must be something peculiar about the climate there, or every one in that neighbourhood could never produce both sheep and wool of such remarkable excellence.   Undoubtedly the climate of Mudgee is favourable to the growth of fine wool, but not more so than that of many parts of Victoria, New South Wales, or even Queensland.   The superiority of these sheep is due more especially to their breeding, for just as Mr. McArthur laid the foundation of the Australian Merino, so the Messrs. Cox laid the foundation of the Mudgee flocks. These gentlemen have held their Mudgee runs for the last fifty years.   In the first instance, Mr. Wm. Cox, the elder, had, as well as Mr. McArthur, a portion of the Spanish Merinos which were imported from the Cape of Good Hope ; then Messrs. George and Henry Cox followed in the footsteps of their father ; and, lastly, Mr. Edward King Cox surpassed both his father and brothers in the improvement of the Australian Merino.   In fact, Mr. E. K. Cox may be considered as second only to Mr. McArthur; the latter being the creator and the former the improver of that admirable race of sheep. The reason why the Mudgee sheep and wool continue to excel is, simply, that these superior animals are there easily procurable, whilst to drive a mob of rams four or five hundred miles is both perilous and expensive. Moreover, the neighbours are sure to get the preference; and should the noted ram-breeders of Mudgee have any extra good rams for sale, we may rest assured that a sale will be readily effected there, where their merits are matter of notoriety. In all probability, for instance, Mr. Baillie can get from his neighbours better prices for his rams than any of our northern squatters would give. Notwithstanding this, it is a mistake to assert that all the sheep in the neighbourhood of Mudgee are good. It would take the climate a long time, unassisted by cash and skilful management, to improve such sheep as those of Mr. Bloodsworth, Mr. Tindal, Mr. Fitzgerald, or Mr. Bowman, to an extent which would render them worthy to compete with the flocks of Mr. Baillie, Mr. Cox, or Mr. Riley.

In point of climate adapted for the growth of fine wool, nearly the entire western and south-western districts of Victoria are quite equal, if not superior, to Mudgee. The Darling Downs, also, I consider to be a country where first-class rams may be bred with profit. This may appear doubtful to some, who regard only the extreme northern latitude of that fine district; but this is more than counterbalanced by its altitude, which must be over two thousand feet above the sea-level. Anywhere on the Condamine, above its north branch, good rams—in constitution, carcase, and wool—might be bred. It is a general and not unreasonable impression that a professed ram-breeder should never be a large general sheep-owner, because it is only natural to conclude that, before he sells, he will take care to select the very tops or best of his rams for his own use; a proceeding which of course gives anything but satisfaction to purchasers. For his stud-flock, this course is perfectly justifiable and correct, as in that case it is apparent that owner or breeder and purchaser are alike benefitted; the former by the enhanced price which he will secure, the latter by the superiority of the rams he will obtain. Should the ram-breeder be also an extensive general sheep-owner, possessing say some thirty or forty thousand ewes in his ordinary flocks, he will yearly require, for their service, a large number of his best rams, because he is bound to keep up the character of the wool from his general flocks, lest the clip, fetching in the market no better average price than that of his neighbours, these latter should say, " Why, our wool is plainly as good as yours, consequently our sheep are equal to yours, therefore in future we will breed our own rams." This, however, although a common view to take of the matter, is not an entirely correct one, for the price of wool, per ℔., is not always a proof that the production of it has remunerated the grower, as a first-class combing or clothing often pays better than the super-combing or clothing ; because, generally speaking, the former is much heavier per fleece than the latter, and sheep-farming must be regarded, like all other pursuits, as a matter of pounds, shillings, and pence.

To obtain a good price regularly, it is very essential that your brand should be well and favourably known, not only to the salesman and broker, but also to the manufacturer, to whom it is of the utmost importance that he should know the brand of the raw material he has to work up. Two wools of equal fineness may differ very much in their intrinsic value to the manufacturer, especially as regards the evenness of the fleece, that is, according to the per centage of superior wool found in each. No manufacturer can arrive at a perfectly true estimate, unless he has had the wool on a previous occasion ; and when a manufacturer has once or twice bought a brand that has given him satisfaction, he will give 2d. or perhaps 3d. per ℔. more for it than any other buyer in the saleroom. All fleeces defective in any respect should be put in bales by themselves. I have often heard it remarked that the good wool would sell the bad : a shallow notion and a grave mistake : the bad will materially depreciate the value and character of the good. As already shown, much depends upon your wool being favourably known ; that it be well classed, carefully skirted, and faithfully described, that is to say, all the fleeces in each bale should be of an even character, and as nearly as possible of the precise quality represented by the marks or brands on the bale. I have often noticed bales marked " second" or " third clothing." Sheep producing this kind of wool should never be kept, nor even a third class " combing," except by farmers who feed their sheep upon English grasses, lucerne, etc. It might perhaps pay such persons to keep sheep of this class for carcase only, but no large sheep-holder, on the ordinary grass lands of this country, should either breed or keep such sheep. A " second combing" you are often compelled to make, as the variableness of the seasons will often cause a vast difference in the fleece of the sheep, and a good " second combing" wool, if sound, will often bring a satisfactory price.

CHAPTER IV.

ON SHEEP'S TEETH.

ALL the British writers* on this subject, whose works have
come under my notice, agree in fixing five years as the age at
which the sheep, in technical phrase, "fills its mouth," that is
to say, acquires eight fully developed incisors ; but in Australia
I have usually found a sheep's mouth to be "full" at four years
of age or a little after. After that period, until the animal be
about six or seven years old, a pretty correct estimate of his
age may be formed, by examining the two last or outer teeth
on each side of the mouth. These teeth rise only when the
sheep is three or four years old, consequently they will not have
had so much work to perform as those which have been in use
already, since the animal was about fifteen months old, and
thus they will be found to preserve their sharp edge, and be
but little worn down when he has reached the age of six or
seven. The condition of a sheep's teeth depends, I may say,
entirely upon the constitution of the soil of the country on which
he was bred and fed ; and taking into account the age also,
this condition will present grades of difference amazingly
numerous and minute. It is not uncommon, or rather it *was*
not uncommon before overstocking came so much into vogue,
to meet with sheep which, from their lambhood, had always
had an abundance of soft, long grass, still at ten or twelve
years of age possessing the incisor teeth more perfect than
those of other sheep, at four years old, which had been fed
upon an overstocked run. Nothing short of a file or a grindstone
would wear down a sheep's teeth like grazing on an overstocked
run, especially if the soil be sandy. This is a matter which
has attracted less attention than it assuredly deserves; I will
therefore endeavour to show its great importance, by illustration.

On a run where the grass is three or four inches long, a
sheep, at one single cut of the incisors, will obtain a whole
mouthful. This is immediately thrown back to the molars,
and by them prepared for the first stomach. These molar teeth
are protected by perfect masses of enamel, and are never known

* "Loudon's Encyclopædia," Part III., Book VII., Sec. 6413; "Youatt on
Sheep," p. 5, ed. 1864.

to fail; so that the importance of promptly relieving the incisors, and throwing the attrition upon the molars is manifest. On an overstocked run the very reverse takes place. The sheep has to make 10 or perhaps 20 bites before he can pass back to the molars as much food as, in one bite on a well-grassed country, he transfers to them. In the former case, the sheep secures, at one bite, some three inches of tender and succulent food; in the latter, he gets, at the expense of much greater time and exertion, and with gravely augmented wear and tear of the incisors, perhaps a similar quantity of dry, wiry, and almost innutritious fibre; for grass is, as all know, much drier, harder, and tougher at or near the "tussock" than in the extended blade. But this is not the worst of it; the short grass is more fibrous and tenacious, so that often it is not severed by the bite, but twitched up by the roots, bringing with it into the animal's mouth more or less sharp, keen sand; and here we have the grindstone, owing to the ruminant nature of the creature, in full and constant operation. A few months since, I was employed by an influential squatter to deliver a flock of ewes, none of which could possibly be more than six years old, and only a very small per centage of them had reached that age. Yet when they were "mouthed," at the time of delivery, fully 25 per cent. of them were found with teeth worn down to the gums, and no less than 50 per cent. were actually broken-mouthed. As before observed, none of these sheep could by any possibility be over six, and many no more than four or five years old, because the run on which they were bred had been converted from a cattle to a sheep station only six years before the transaction adverted to; no ewes were ever placed on the run, excepting those from which these sold ewes were bred, and they bore a different ear-mark from that borne by the station-bred sheep. The purchaser, however, refused to take delivery, believing them to be older than they were described to be by the vendor, until he was shown a lot of six-tooth sheep, whose teeth, the four first incisors, were worn comparatively as close down to the gums as were those of the eight-tooth ewes he was purchasing.

It is often said that short grass is the sweetest and best for sheep, but the assertion, in itself of very doubtful authority, must be limited by a consideration of the relative lengths of grass. If the grass be so long that the animal grazes without taking up particles of earth or sand, it is no longer short grass within my meaning; if it be shorter than this, then all the mischievous effect upon the incisors, to which I have adverted, will be experienced. I once saw a flock of four-toothed sheep with teeth completely worn down to the gums, and the two lamb's teeth on each side preventing the mouth from closing on the grass; the sheep were, of course, wretchedly poor; yet they had been fed on long grass, and the defective state of their teeth arose from their being infested with fluke, which caused them, prompted by instinct, to seek, by licking and even gnawing the ground, to obtain salt. Sheep thus affected are technically called "dirt-eaters;" but the subject will be found more amply discussed in my chapter on "Fluke."

Great differences of opinion exist as to what constitutes "broken-mouth" in sheep; the buyer and the seller, indeed, very rarely agree upon the point. As, during the past fifteen years, I have been engaged in purchasing largely—first for the late Mr. James Maiden, afterwards for Mr. Hugh Glass, and for the constituents of J. H. Clough and Co.,—especially the description of sheep termed full-mouthed, and as I have thus purchased in one year as many as 48,000 or 50,000, all of which I mouthed, I trust I shall not be deemed egotistical if I claim to be considered an authority on the subject. Being fully aware of the conflicting interests between buyer and seller, and the consequently widely divergent views taken by each as to what really constitutes a broken-mouth, I invariably inserted a clause in my agreements, constituting myself the sole judge of what should be considered a broken-mouth, without any reference whatever to the seller. This clause I emphatically pointed out to the seller before the purchase, and I never, upon any occasion, had any difficulty in thus carrying out my arrangements amicably. True, some people have objected to sell to me under this agreement, but this occurred only when the owner knew that the sheep would not bear inspection. The sheep I consider broken-mouthed may be thus enumerated: First, any whose teeth are worn to a point; second, any whose teeth are divided, that is, which do not touch each other at the sides, from top to bottom, from crown to gum; thirdly, any having teeth worn down close to the gums; fourthly, any whose teeth are not perpendicular, where the top part of the teeth leans forward, so as to strike the pad of the upper jaw on the outer edge, or perhaps over it; fifthly, any whose teeth exhibit an unequal cutting surface, that is, which are worn down in one place more than in another. In fact, I consider that any sheep whose mouth is so imperfect as to prevent his feeding with perfect ease and effect should be placed in the category of the "broken-mouthed," whatever be its age. With special reference to that kind of broken-mouth enumerated as "fourthly," I must remark that, upon under-stocked runs, or runs where long, soft grasses abounded, I have seen sheep twelve or fourteen years old with mouths quite perfect, except that the teeth were pushed forward to a degree which would prevent their striking the pad. This is invariably a sure sign of very great age; the teeth may be at the same time all firm, close, and even, and, in fact, nothing to denote old age, except this want of perpendicularity, and the teeth being somewhat longer than ordinary.

Heathery runs and runs bearing many different kinds of shrubs, especially of the acacia tribe, of which sheep are very fond, are, from the fibrous nature of such herbage, highly destructive to the teeth. Fortunately, most of the salt-bush is of a softer texture, and does not point and divide the teeth as do the more tough and fibrous plants.

At from fourteen to sixteen months, the two central milk or sucking teeth fall out, and are replaced by two sheep's, or permanent teeth; but this is not invariable, as I have often known sheep to be twenty months old before the first two permanent incisors were up; whilst in other cases it is not unusual to find the animal at that age with four sheep's teeth. At two years, nearly the whole flock will have four teeth, and at from two years and nine months to three years, they have their six teeth well developed. Then, again, at from three years and nine months to four years, they will acquire their two last teeth, when the mouth will be what is technically termed "full." Of course, every sheep may not fill its mouth at the exact period I have named; some may be earlier, others a little later, but as a rule, the time I have indicated will be found correct for all practical purposes. The reasons why some sheep get their permanent teeth earlier than others I cannot absolutely determine, but some persons have not hesitated to attribute the circumstance to good feed and condition. With this view I cannot concur, for my own experience has led me to believe that a lamb which has been stinted of milk up to the time it was weaned, or one that has been weaned too early, will get its teeth before the well-fed one. The reasons for this are, I submit, first, that the lamb, being deprived of a portion of the sustenance which it ought to derive from the mother, would necessarily be compelled to make up the deficiency by consuming an extra quantity of grass, which would, as a matter of course, wear away its teeth much sooner than those of the one that had an abundant supply of milk, and therefore was not obliged to subject its teeth to much wear and tear whilst they were so delicate and unfitted to sustain it. Secondly, I think it highly probable that the premature wearing away and removal of the lamb-teeth induces ever-helpful Nature to come to the rescue, and stimulate the permanent teeth into exceptionally early development. Many of my more experienced readers will call to mind that they have observed occasionally some very small sheep amongst their weaners, and upon catching them they have found that these "poddies" had all two teeth up, whilst the bulk of the flock had not even commenced to show signs of cutting their incisors.

I should, perhaps, not have extended my remarks on sheep's teeth to what may be deemed an unconscionable length, but for my earnest desire more fully to demonstrate the grievous error of over-stocking. I have elsewhere shown* that this irrational system tends greatly to reduce the size of the carcase, that it deteriorates the quality and diminishes essentially the quantity of the wool; but here we lose the sheep entirely —for such abortions are absolutely valueless—before it can arrive at its miserable, stunted maturity, in consequence solely of the terrible hard work we compel it to perform in order to eke out its mere existence.

* *Vide* chapter on "Over-Stocking."

CHAPTER V.

ON CULLING EWES AND RAMS.

To class sheep properly, requires an amount of skill which nothing but extended practical experience can confer. There are many defects in sheep, both of carcase and wool, which cannot be clearly defined to the apprehension of the wholly inexperienced person, by any description in writing. I will, however, endeavour to point out, for the benefit of the tyro, some of the more glaring imperfections. If the sheep-owner has not an opportunity of getting his sheep classed by an expert, he may, by observing these instructions, be enabled to detect and throw out of his flocks such sheep as are palpably bad. Sheep of the following descriptions may safely be rejected:—1. A very short-stapled, hard-woolled sheep. 2. Sheep that strip at the points, and lose the belly-wool, having a clean head without topknot. 3. Any that have black or yellow spots on legs or face. 4. Unusually small sheep. 5. Any that appear thin, and constitutionally feeble. 6. Sheep whose wool is thin and light. 7. Those which have short white hairs on the face, under the arm, and inside the thigh. 8. Any with very coarse wool about the breech and tail. 9. Any with long hairs appearing on the surface of the fleece. 10. Any sheep whose wool, at shearing time, is less than one inch in length on the ribs or wither. 11. Any long-legged, small-bodied sheep. 12. Sheep dipped in the back, or otherwise mis-shapen. It is not unusual to hear people descanting on the merits of some particular animal, and winding up by declaring that the wool on him "opened like a book." Either the phrase or the idea is incorrect. Wool, when you open it on the sheep, should not open in long layers; on the contrary, no part of the fleece should open but that which you actually touch, and no disturbance should be apparent at a distance of three or four inches from your hand. In a really good fleece, every fibre of wool should grow independent of any other, like hairs in a broom, and no two should be entangled with each other, from root to point. Sound wool of this kind would comb well, if no more than an inch or an inch and a half in length. By carefully culling such defective animals out of your flocks yearly, you will find both the quantity and quality of your wool improve rapidly, and it is more than useless to keep them, since their progeny will most assuredly be worse than themselves. It does not require any great amount of skill, nor even of common sense, to detect these glaring imperfections, yet if these brief hints be carefully and intelligently put in practice, little will be left for the classer to do, and whilst doing this you will soon be enabled to detect other faults, which I shall endeavour to make as plain as possible when treating of defective wools. It is often said that the difference between a good sheep and a bad one is mere matter of opinion, since the qualities sought in sheep by one classer are entirely ignored by another. Every classer has, I doubt not, a certain standard of perfection always in his mind's eye, and there may be, also, certain prejudices against and weaknesses for sheep of special characteristics ; as, for instance, with some classers nothing will compensate for a want of fineness, whilst with others length of staple is the *sine quâ non*. I would not reject a sheep because the wool fell off a little in fineness, provided its qualities were satisfactory in every other respect ; that is, if it possessed a fleece of moderate length, softness, evenness, density, soundness, and free to comb. Fineness is undoubtedly a great consideration, but many very imperfect sheep are kept solely because they possess this distinction in a very eminent degree. But quantity must be considered as well as fineness ; one pound, one and a half, or two pounds of fine wool will not yield the money value of four pounds of a somewhat coarser kind ; so that if a sheep possess all the valuable properties enumerated above, together with a good sized, symmetrically-formed carcase, I should keep him in preference to one whose wool was finer, but wanting the other conditions.

In this place it will be as well to say, in reference to stud-flocks, that although the ewes may to all appearance be everything you may desire as regards form and wool, yet particular attention is requisite at the time of lambing, to observe closely if any of the ewes are bad nurses, that is, to see that the wool of any ewe does not get thin and puny ; to see that she has plenty of milk, and does not get out of condition whilst rearing her lamb. A good sound-consitutioned ewe should rear a strong lamb, and yield a good fleece at the same time. Any ewe that " strips" her belly-wool, or loses her wool at the points whilst rearing a lamb, should be discarded. I would also reject a ewe which throws a lamb with black or yellow marks on face or legs ; I would not absolutely object to the tips of the ears being a little brown, but I would prefer them white.

The neck of the ram should be short and thick, free from straight hairs on the under part; and without dewlap. Horns spiral, not too close to the head; the first curve should be about one inch, or one inch and a half from the cheek; a horn projecting straight from the head is very objectionable. Yet I must admit I have seen very good rams having this defect. All Merino rams should be horned; almost every one seems to have an objection to poley or snail-horned rams, and I share in the prejudice. I have certainly seen excellent rams both of the snail-horned and poley description; yet, on account of the resemblance between the shape of the horn and the curve of the wool, I would prefer a poley to a snail-horn; and as in the poley we are deprived of one useful test of merit in our wools, I should generally reject a poley ram. I have heard it remarked that a horn growing close to the head denotes high breeding; this is observable in Stigers sheep particularly. Nevertheless, it is very inconvenient, and sometimes fatal to the life of the animal, as the horn in growing gradually forces its way into the jaw, when it requires great skill and care in cutting it out at the back. The twist or curve of the horn should be short and regular, as the curve of the horn and that of the wool are mostly found to be proportionate. There should be on the forehead a good topknot coming down to the eyes; the face small, and free from wool, but covered with fine soft down. The eye should be full, mild, and quiet. The carcase should be capacious and round; straight along the back, but not too long; wide across the shoulders and loins. The legs should be short and symmetrically placed. The wool should be from three to four inches long on the best part of the shoulder; and on other parts of the sheep it should approach this standard as near as possible, both in length and fineness. The top or surface of the entire fleece should be level, as though a smoothing plane had been passed over it; it should be thick, but of a dusky blue colour, feel soft to the touch throughout, and possess abundance of "yolk." In handling, the wool should fill your hand, so that you should not be able to close the fingers on to the palm. I have known many buyers of rams, and good judges too, who would not breed from one which had not wool all the way down to the hoof, although in other respects the sheep might be unobjectionable. This I consider a mistake. It does not follow that, because a sheep is well woolled below the hough and knee, the fleece is good on the body. For my part, I prefer a sheep whose wool comes well down the shoulder without shortening, and the same down to the hough, yet without wool below the knee and hough. This, however, is mere matter of opinion, and perhaps worthy of mention only because a fallacious notion is by some persons entertained that the wool on a sheep's legs constitutes a test of good breeding, a notion which experience has not borne out. All sheep, without exception, should be free from black, yellow, or tan spots on the legs or face; and, in the case of rams, no black stripe on the horns, nor yellow or tan colour on the roof of the mouth can be tolerated. I do not admire dewlap, sometimes called throatiness, or leather-neck. Sheep displaying this peculiarity in any extraordinary degree are, it appears to me, usually close between the fore-legs, and, possibly as a result of this construction, not of strong constitution. Certain it is that, amongst wethers, they are commonly the last to fatten, and rarely become very prime.

CHAPTER VI.

ON PUTTING THE RAMS TO EWES.

THE time for coupling varies in different districts, according to temperature and the period when the rains usually set in. The sheep-owner must be guided in this respect either by his own experience or the custom of his neighbours. In New South Wales, in consequence of the great diversities of climate, the period of putting in the rams varies extremely. In the districts of Bathurst, Goulburn, Yass, New England, and wherever a low temperature prevails, the rams are placed with the ewes from about the first to the twentieth of May, so that lambing does not commence until October is well advanced, when the weather is warm and the spring grass abundant. On the lower part of the Lachlan, and throughout the Riverine country, the rams are put to the ewes about the end of November or beginning of December, so that lambing commences in May. To ensure a good lambing, your rams and ewes should be in top condition. I have heard it remarked, " the ewes are too fat to lamb, or even to breed, and a fat ewe does not usually throw a good lamb." From these dicta I entirely differ, for in all my experience I never have seen a ewe too fat either to take the ram or to lamb safely. On the contrary, the fatter your ewes are at lambing time, the better your lambing will be. Some years ago I was on the Darling Downs, and there I saw the fattest ewes I ever saw in my life, with an increase of 98 per cent., and that not on one flock only, but on five flocks of over 2000 each. At weaning time they drafted off 4000 ewes that had reared lambs—culled ewes, that is, inferior woolled ones—and sent them to a boiling-down establishment; they yielded 23lbs. 10oz. of tallow each! This, be it observed, took place within one week after weaning, and there were in the entire lot certainly not more than 3 per cent. of dry ewes. The sheep were well-bred Merinos. Great care should be taken that the rams are shorn six weeks, or even two months, before they are put in to the ewes. Perhaps it may not be generally known that a newly-shorn ram will not beget a lamb. From experience, I know this to be a fact, although I can do no more than conjecture a cause; namely, that the constitutional vigour of the animal is diverted from the procreative faculty, and concentrated in the reproduction of its natural and indispensable clothing; certain it is, that the wool never grows so rapidly as it does immediately after shearing. In the year 1838, we had a very bad season; consequently, a poor lambing ensued. At shearing time, in December, I drafted off all the dry ewes (about 4000), and had them shorn by themselves first; I then selected 100 of the best of the rams, young and vigorous, or 25 rams to each 1000 ewes, had them shorn, and put them to the ewes. I kept them with the ewes for one month—five weeks being the period usually allowed in the Bathurst district—and out of the 4000 ewes, only 165 lambs were dropped. Of course, I was perfectly astonished, and made every possible inquiry upon the subject, but no one could afford me any information. Some three or four years after, whilst looking over a copy of the *Sydney Herald*, I observed a paragraph, stating a case exactly similar to my own, which had occurred to (I think) a Mr. Campbell, in the Goulburn or Yass district. The paragraph stated that this gentleman had drafted his dry ewes and shorn them, and afterwards shorn as many rams as were needed for them; he left them together the usual time, and after all, the ewes did not drop 5 per cent. of lambs. After this, and not without much cogitation, I began to suspect the cause, and determined to test the soundness of my conclusion. For this purpose I drafted, at shearing time, 100 dry ewes, and put to them four young, vigorous rams; the dropping amounted to nine lambs. Am I not justified in concluding that sheep newly shorn will not produce lambs? I would always make it a rule to put my best rams to my best ewes, and I would ensure their superiority in the following manner, viz. : Say you have this year 1000 ewes for the ram; before shearing, you should go through your rams, and select and mark the very best of them, say, to the number of 20. Of these again you find that 10 are superior to the other 10; put these best 10 in first, with half a dozen stags, if you have them, to "tease" the ewes into season; keep these 10 in for three weeks before putting in the other 10, so as to give the best 10 an opportunity of getting as many lambs as possible; add then the other 10, and keep the whole 20 in for the remainder of the time. Many people keep their rams with the ewes during eight or nine weeks; this I think a mistake, as it makes a straggling lambing, and spoils the appearance of the lamb flock by the inequality of their size when weaned.

D

Since paddocking sheep has become prevalent, and especially where the paddocks are large, the ewes too frequently are found to split into small mobs, and so, perhaps, you will find 50 rams with the same number of ewes, and the great body of the sheep without a ram at all. To remedy this serious evil, a yard should be constructed somewhere about the centre of the paddock, where the entire lot contained in the paddock should be yarded three 'nights every week, or oftener, should it be found that the ewes split into mobs. Obviously it is essential that the rams, during the period allotted for impregnation, should be in the society of the ewes as much as possible; indeed, very considerable numbers of ewes would not come into season at all unless "teased" into that state by the pertinacious solicitings of the rams. Yet another good reason why the ewes should be yarded occasionally is, that a single ewe will often decoy a ram away by herself. I have seen, hundreds of times, a single pair thus drop back from the flock, and before they were found or rejoined the flock, the ewe would perhaps have been tupped fifty times, and the ram exhausted and unfit to beget another lamb during the season. This could not occur if the sheep were yarded every time they were found in small mobs. Again, by thus keeping the sheep together, the ewes would come more quickly into season, and the period of lambing would be materially abbreviated, so that the risk of losing or stinting the younger lambs at the time of shearing would be greatly lessened. Six weeks, provided a sufficient number of vigorous rams, in good condition, are employed, I consider quite sufficient time to leave the rams with the ewes. It is the common practice to allot two rams to each one hundred ewes. This number is certainly sufficient, provided the rams be young, vigorous, and in good condition, but some people put more; in fact, they put in all the rubbish they possess, on the principle that it is better to have a bad lamb than no lamb at all. Nevertheless, it would be much more sensible, because more profitable, to put only three rams to 200 ewes, provided always the rams were of good quality, young, vigorous, and in top condition. We have often heard it asserted that a single ram has begotten nearly 100 lambs in a single night; and putting one and a half rams to beget 100 lambs in six weeks cannot certainly be to overtax their powers. By way of illustration, a mode of reasoning to which my readers will perceive I am prone, I will relate a fact which came directly within my own experience. Mr. Lawson, of Macquarie Plains, about the year 1836, was sending up from that station to Davy's Plains a very superior ram, and I borrowed him for one night as he passed the station I had charge of.   I put him into a flock of maiden ewes at about five in the evening, and took him out again about seven o'clock next morning, and at the proper time these ewes dropped 39 lambs, within two days, or 48 hours.   If three rams only are allotted to 200 ewes, they must be young, say four or six tooth, at which age rams are usually most vigorous.   Care should be taken that none are lame, especially in the hind legs or quarters.   Rams will tup at a very early age, so early as four or five months; indeed, I have myself known ewes to be tupped by their own lambs of less than five months old, and prove in lamb to them.

CHAPTER VII.

ON LAMBING.

To ensure a good lambing, three conditions are indispensable, viz,, plenty of feed, an ample supply of water, and paddocks of dimensions sufficient, but no more, for the number of ewes to be lambed. Nothing has been more detrimental to lambing, and the production of good, well-grown lambs, than the use of large paddocks. Nor should ewes, if it can be avoided, be lambed down in large flocks. I call 6,000 or 7,000 a large number to lamb in one paddock, however extensive. It is, however, less injurious if the lambs are dropped in March or April, as then they will have acquired sufficient size and strength to be weaned before washing and shearing. A ewe should never be washed and shorn until her lamb is weaned from her, unless the sheep are in small flocks, so that one day's washing, or one day's shearing, or at most two days' washing, or two days' shearing will finish a flock. The winter rains for a few years past have occurred so late in the season that many sheep-owners have been compelled to lamb in June or July, there not being grass sufficient for the ewes and lambs at an earlier period; in fact, the owners have been compelled to bring on the lambing at a late and the shearing at an early period, and the consequence is that half the lambs get weaned before they are old enough, become " poddies," and never attain the size or symmetry of a full-grown and well-developed sheep. Take, for instance, an ordinary sized paddock, containing 6,000 ewes and their lambs : say they wash 1500 per diem, four days will be occupied in washing the ewes, providing Sunday or bad weather does not intervene. Three days must be allowed for drying; on the fourth day they commence to shear the first day's washing, on the fifth, the second day's washing; it will take seven consecutive days to finish them ; there is also one Sunday, and not less than one day will be spent in taking them from the paddock to the wash-pool, another in taking them back ; making in all ten days; that is, providing there occur no break in the weather, and no mistake in the general management of the station. Now, how many lambs have been away from their mothers all this time? doubtless a very considerable per centage. Some people will say, " Oh, yes, this is all very true, but we do not manage sheep on our station in that foolish way." The proof of this assertion can be had only when the ewes come to be turned-up on the shearing-floor, and I solemnly affirm that of the scores of shearing sheds I have been in, I never saw one where this unpardonable mistake was not plainly apparent, from the distended and inflamed udders of some of the ewes. In South Australia, Victoria, and New South Wales, I have carefully scrutinised this matter, and in each colony I invariably found the same practice. In South Australia, however, the least so, and solely because the paddocks there are of less area. If such an arrangement be at all possible, not more than 1,500 to 2,000 ewes should be permitted to lamb in any one paddock, unless (as I before observed) they lamb in March or April, which, as the ewes are generally shorn last, would give the lambs five months before weaning, August being now the favourite month to commence shearing. It would be much better to shear in the grease ewes that do not lamb until June or July, or even, if it can be so managed, not to shear them until November. Lambs, when weaned, should, if possible, be kept in the paddock they were dropped in; for if put into a strange paddock, especially if watered by means of troughs, their shyness will cause them to suffer in condition from want of water. If at all practicable, they should be placed in a paddock with water-frontage, or dams, and a few old sheep turned among them, to steady them and lead them to water. It is easy to keep lambs in condition, but if you allow them to fall off and become poor, you will find no sheep so long in recovering themselves; in fact, it is doubtful if they ever perfectly get over it. Besides, a very slight check upon them will surely cause a break in the wool; but on this point I purpose writing more fully when treating of defective wools.

When ewes are lambing, they should be left as quiet as possible; there is no greater mistake committed, than to be continually riding through the flock with a dog after you. At the period of lambing, the natural timidity of the animal is greatly exaggerated by the strong maternal instinct, and she suffers, when thus disturbed, paroxysms of fear and anger, which cannot but be most injurious to herself and her offspring. It is very rarely that a healthy ewe actually requires assistance in the time of parturition, and probably nine times out of ten it were better to lose one than to disturb the whole flock, especially since it is a chance if you could do her any good when you have caught her. A man on foot could stroll gently among them occasionally; he might have a horse with him, but when he draws near to the sheep, he should hang-up, or otherwise secure his horse, walk very quietly up to them, and by no means follow them, should they string off.

It is my impression that more harm than good is done by
going near them at all, unless eagle-hawks, crows, or dogs, are
troublesome. In many cases it would be quite sufficient to
ride round the outside of the fence, and see that it was
perfectly lamb-proof and secure. Neither should lambs be
taken out of their own paddock to be cut; the yard I pro-
pose to occasionally put the ewes in when the rams are
with them, would answer for this purpose also, with just a
division fence run up the centre, and a few saplings or hurdles
to make a race; the lambs could be run into a small yard,
operated upon, and then put among the ewes that had already
passed through the race. When all the lambs in are cut, you
will of course count your ewes, and should you find that any
considerable number are still away, hold those you have in
hand until you complete your muster; and when you have
done, open wide your hurdles, and let them draw-off of themselves. This is certainly better than driving them perhaps
eight or ten miles to the usual permanent drafting-yards, and
there keeping them for forty-eight hours without food, as is
only too frequently the case. Nothing, be assured, is more
prejudicial to a sheep than long fasting, yet it is quite common
to see the poor animals confined in yards, without food or
water, for two, or even three, days at a time. I have often
heard people say, "Oh, another night cannot harm them,"
when they have already been twenty-four hours in the yard;
but I am perfectly certain that such inhuman and irrational
treatment not only throws the sheep back in condition, but
eventually injures seriously its constitution. Some persons
draft off their dry ewes, together with such as have lost their
lambs, and again put the rams among them. I do not think
the practice a profitable one, it will throw them completely
out of season, and the lambs will be a perfect nuisance till
weaned. I may add here, that I never saw a fat or well-
conditioned ewe deficient in milk, whilst among starved ewes
I have seen thousands without sufficient milk to rear a kitten.

In conclusion, I submit to my readers a plan which I have
seen adopted with advantage, for getting together the ewes and
the lambs belonging to them, in a large paddock of say 6,000
ewes. The day before the sheep are required to go to the wash-
pool, let the paddock be clean mustered, and the sheep placed
in one compartment of the yard, at the centre of the paddock,
which I have recommended in a preceding part of this chapter.
At two or three points of the central dividing fence, let the
brush or hurdles be previously removed, and their place
supplied by good strong stakes, driven firmly into the ground,
and well secured at the top, so that they shall resist a heavy

pressure in whatsoever direction applied. Between every two stakes let a space be left sufficient for the passage of a lamb, but not for that of a ewe. When these "lamb-filters," as I will call them, are not required, the stakes can be easily masked with hurdles, which for this purpose are handier than brush. Let your muster be completed by five or six o'clock in the evening. Now run through your race into the other and hitherto empty compartment of the yard, the number of ewes you want to wash next day, and leave them there all night, having removed the mask of hurdles from your several "lamb-filters;" when morning breaks, you will find that all the lambs belonging to the drafted ewes have passed through the filters, and rejoined their mothers. One or two exceptionally large lambs may have been unable to pass through the filter, but in this case you will find the mother and the lamb close together, although on opposite sides of the dividing fence. Repeat the operation each night, only, of course, as each day's mob is sent to the wash-pool, the remainder must be shepherded, until the last day's washing has left the yard, by which time the first day's washing will have been shorn, and you will have a clean paddock to put them in. This plan is not novel, although little used of late; but if it be carefully worked by good men, under good superintendence, it will be found most advantageous.

As immediately connected with the subject of Lambing, it is proper in this place to mention a disease affecting lambing ewes, and known as "garget." It consists in an inflammatory swelling of the udder and teats, arising from excessive secretion of milk, in consequence of which the teats become too large for the lamb's mouth. The udder first assumes a dark red colour, and shiny appearance, then it becomes purple, and lastly it becomes gangrenous, and mortifies. The disorder occurs most frequently when the ewes are in high condition; but if proper care and watchfulness be exercised, it may be easily detected at its commencement, when it is as easily remedied. About 1846–7, I had occasion to deal with many cases of "garget," and feeling dissatisfied with the treatment then customary, viz., hot-water fomentation, and occasionally milking the ewe, I sought the advice of a very clever medical gentleman, Dr. McHattie, of Bathurst. The doctor recommended me to try local bleeding, from the large belly-vein, as near as possible to the udder, by making in that vessel a longitudinal incision of about three-quarters of an inch in length, and leaving it to bleed as long as it would. I then feared, as many persons will now, that the ewe would bleed to death, unless the orifice were closed with a pin, etc., as in ordinary cases of phlebotomy, but I never knew this to occur; and out of some hundreds upon which I have operated, provided the disease was not too far advanced, and the blood flowed readily and freely, not a single failure ensued. On first opening the vein, the gush of blood is violent, but as depletion proceeds, it gradually slackens, until the air dries the oozing, and the wound closes and heals quickly. The lambs of very fat ewes, it should be here observed, are, when first dropped, somewhat small, which, for obvious reasons, seems to be a wise provision of nature; but observe them when they are six or eight weeks old, and you will find them half as large as their mothers. Most people have seen pet ewes about a head station, basking under the verandah, and scarcely able to crawl for fat. Who ever heard of one of these sheep dying from excessive fatness whilst lambing? No; but it is often remarked, "What a very small lamb Jenny or Peggy has!" But look at Jenny's or Peggy's lamb when it is five months old—that is, when it is weanable—and see if the same fault can then be found. The question of high or low condition for yeaning ewes is still, however, a moot point amongst sheep farmers, and I claim only the privilege, which most willingly I concede to others, of stating my opinion, which is decidedly in favour of the high-conditioned animal; indeed, I may say that I never saw ewes too fat to lamb with safety.

Flockmasters are divided in opinion as to the advantage or disadvantage of their ewes dropping twin-lambs. By one section it is considered better to have one good lamb than two small ones; others desire a rapid increase of numbers, and like to find their ewes dropping twins. There can be no question that single births do always produce a larger and stronger lamb than twins are ever found to be; and very few Merino ewes are capable of successfully rearing two lambs. There are, of course, exceptional cases where 'a Merino ewe has dropped and reared two average lambs; still, I give in my adherence to the single lamb side, because the strain upon the constitution of the ewe is too great; her condition is more than usually reduced, her wool becomes mushey and tender, and it loses considerably in weight. Many persons, however, hold that to breed twins is profitable, and they conceive that the best way to secure this increase is to keep for breeding all lambs that have been dropped twins, as they consider the peculiarity to be hereditary. Very possibly this may be the case, but the climate has much influence also. Now, in all the countries I have visited, with one exception, the average dropping of twins, from ewes of three years old and upwards, is from four to 12 per cent.; in young ewes, that is, ewes with their first lambs, about three or four per cent. is the usual thing. Take the two higher numbers—12 and four—and we get a mean of eight per cent. all through the lambing flocks. This, I think, will be found to be about the general average, except on the Darling River and its vicinity. There it is not uncommon to have 30 or even 40 per cent. all through the lambing flocks. I once saw, in that district, 27 ewes that had dropped their lambs on the first day or night of lambing, and 17 of them had dropped twins. I am not aware that any person has attempted to account for this singularity of the Darling country. Mr. Sadlier, the manager of the Albemarle stations, informed me that frequently he had cut 120 to 130 per cent. of lambs, and from the dropping to the time of cutting, many of the twins must have died, as they are more liable to lose the run of their mothers than when a ewe has one lamb only. If a ewe finds she has one lamb at her side, she will not bleat and run about in search of the other, until she has, by some weeks of maternal care, learned to look for both. I may add that the sheep here referred to were all Merinos; indeed, I am not aware of a single flock on the Darling which has been crossed with the Leicester or any of the English coarse breeds. I have sent ewes to the Darling from both South Australia and Victoria that never dropped more than the usual average of lambs in those colonies, yet, when they had been on the Darling a year or so, they dropped twins in numbers equal to the Darling bred sheep. I have also bred from ewes, both in Victoria and New South Wales, which were lambed on the Darling, and had there bred the usual Darling number of lambs, remaining there until they were five or six years old, yet when they came to lamb in Victoria or New South Wales, they produced no more lambs than was usual on the station. Now, what can be the cause of this? Certainly not high condition, for although sheep on that river are usually very good in that respect, yet I have seen sheep in many other districts quite equal to them. In many parts of New South Wales and Queensland I have known the sheep to be mud-fat all the year round, yet I never saw or heard of so many twins. Although we cannot account for this very extraordinary circumstance, yet I think we shall be justified in assuming, from what is stated above, that it is not due to any peculiarity of breed, nor to any superiority in point of condition. On the Darling, and in the "far north" of Adelaide, South Australia, are the worst specimens of sheep to be met with throughout the colonies. Their frames are badly formed, they are leggy, their wool is dry, harsh, light, and almost yokeless, having, indeed, every fault of which wool is susceptible; yet they readily take on fat, and are more than commonly prolific.

CHAPTER VIII.

ON THE VARIOUS KINDS OF COUNTRY USED AS SHEEP RUNS, AND THE DISEASES PREVALENT IN EACH.

A SLIGHT topographical sketch of the various zones or belts of country where sheep are fed, may not be unwelcome to persons intending to embark in pastoral pursuits. The main range of the continent of Australia, called the Australian Alps or Dividing Range, runs north and south. The waters flowing from the western flank of the chain are called the Murray waters, in consequence of their emptying themselves into that river or its tributaries; on the eastern side the waters flow more directly into the sea. On the western side are found the countries best adapted for sheep-husbandry, on the eastern side there is very little good sheep country; none, that I am aware of, in New South Wales; the Richmond and the Clarence have both been tried and failed, as also the Brisbane in Queensland. It is in these eastern tracts that a disease prevails called "gut-rot," which I have elsewhere spoken of. Proceeding south, we find a similar disorder prevalent, but called the "coast disease," as about Western Port, and thence to the neighbourhood of Mount Gambier and the Glenelg. The only really sound country I am acquainted with on the eastern flank of the Dividing Range is Fiery Creek, and a considerable area of the western parts of Victoria; unless, indeed, we may include that part of the Adelaide country about Clare, and the Burra Burra, as Bungaree, White Park, Hill River, Bundaleer, Booborowee, and a few stations in their immediate vicinity. All these are excellent sheep-stations, but they occupy an area so limited that they must be considered as exceptions only to the general rule that the eastern waters as a whole have very few good sheep-runs on them. Turning next our attention to the table-lands on the western flank, we find the whole of that tract more or less unsound and flukey. It comprises the entire table-land of New England. Commencing at the head of the Condamine, and running south to the extreme end of Gipps Land, including the Severn, McIntyre, Big River or Gwydir, Namoi, Peel, Lachlan, Macquarie, Murrumbidgee, Murray, and Goulburn; and doubtless at the head of the Glenelg, fluke will some day be found. In the same category must be placed Bathurst, Goulburn, Manero, Limestone Plains, and all the eastern or upper parts of the rivers. The grasses on this style of country are generally coarse, of few varieties, and not at all nutritious; and they are perfectly unfit to permanently keep sheep upon; the summer months, say from November to March, are in fact the only portions of the year when sheep can be safely fed on these countries. The next country to speak of is the intermediate, being situate between the western flank and the salt-bush. This tract is unfortunately limited, but it is perhaps the best wool-growing country of the whole, possessing at the same time in a great measure the advantage of a salt-bush country, as there sheep are found to require no salt. The grasses are highly nutritive, of numerous varieties, and thick in the sward, there is but little dust, and the temperature does not range high. Sheep may not fatten here quite so quickly as on the salt-bush, but fluke is unknown, so that the herbage must be partly saline, as the ground affords a sufficiency of salt. Next comes the true salt-bush, perhaps the best fattening country of all. It is generally supposed that the hot, salt-bush country cannot grow good wool; it has not yet had a fair trial. The sheep-holders generally in this country have, as a rule, had runs too large; and in place of classing and culling out all their defective sheep, they have been breeding from trashy animals, for the purpose of "stocking-up" their spare country. No class of sheep-owners enjoy so many advantages as these gentlemen possess; they have an almost boundless scope of country of the most fattening kind, at a rent scarcely more than nominal; their sheep require little management, on account of their generally healthy state, and their expenses are at a minimum. Not only is indigenous disease almost unknown, but the salt-bush, as elsewhere remarked, is a cure for both fluke and catarrh. The only tract of salt-bush that I ever knew to be subject to any disease is a portion of the Liverpool Plains district, situate between the Woolshed on the Namoi, and Bondi; a few stations in that vicinity have been subject to the Cumberland disease. From all this it may be inferred that of the four kinds of country enumerated, two only are entirely unexceptionable, viz., that which I have termed the intermediate, and the salt-bush itself.

CHAPTER IX.

ON SHEEP RETAINING THEIR WOOL.

MANY discussions have taken place, as well in Australia as in Europe, upon the questions of how long the sheep will retain its fleece without shedding it; whether any one breed will retain it longer than another; and which breed excels in this respect. From all I can gather by searching the works of highest European authorities on the subject, two, or at most three years, is the extent of the period during which a sheep of any breed, sex, or age, can be expected to retain its fleece without shedding.* In support of my opinion, that a really well-bred Australian Merino will carry its wool without shedding for an indefinite period—nay, for as long as it shall live, under certain circumstances and conditions—I will first state, generally, what I conceive those conditions should be, and then cite some authentic facts in corroboration. Let a sheep, ewe, wether, or ram (if the former, not to be bred from), be placed in a paddock, where the temperature would be moderate throughout the year; let its feed be abundant at all seasons, water easy of access, and it is my impression that, unless overtaken by accident or disease, the animal, if it lived for twenty years, would still, and perfectly, retain its wool. In the year 1863, two bales of greasy wool, in the fleece, were consigned for sale, from New Zealand, to J. H. Clough and Co., the eminent wool-brokers of Melbourne. Some of this wool had attained a length of $22\frac{1}{2}$ inches, and very few fleeces were less than from 14 to 18 inches in length of staple. This wool was, beyond all question, a first-class Australian Merino combing wool, as full of elasticity as curled hair, perfectly sound, without a break in it, and each fleece entire, as I ascertained by spreading them out on the floor, in order, by Mr. Clough's permission, to take samples. Now, the question is, to what age must a sheep have attained, to bear an entire fleece of wool, $22\frac{1}{2}$ inches long in the staple? Let us assume, that within the first year the wool grew three inches, the second year $2\frac{1}{2}$ inches, the third $2\frac{1}{2}$ inches, and the fourth $2\frac{1}{2}$ inches, or $10\frac{1}{2}$ inches during the

---

* I am only recently informed that the result of some experiments on the Rambouillet sheep, in France, would seem to indicate, in that particular case, a longer period. (Youatt, p. 63, ed. 1864).

first four years; we have still, however, 12 inches to account for, which we will do by allowing two inches per annum for the remaining six years, which will bring the age of the sheep up to ten years when shorn. But the most remarkable part of the affair, is the thorough soundness of the wool; it was strong and tense as a fiddle-string, whilst its elasticity was wholly unimpaired. The whole of these facts can be attested by numbers of gentlemen, to whom I exhibited samples of the wool; also, by Mr. Goodrich, woolsorter, at J. H. Clough and Co's. warehouse in Hamilton; by Mr. John Evans, foreman in Messrs. Clough's Melbourne warehouse; and by Mr. Harris, manager for the firm at Hamilton.

At the Victorian Valley Station, near Hamilton, an exceedingly rough country, where it is difficult exhaustively to muster, they have, on several occasions, got sheep in, which had never been shorn, although full-mouthed, and their wool was found quite perfect, without a symptom of shedding, although 12 or 14 inches long. Again, on the Upper Goulburn, it is not uncommon for sheep to come in with four or five years fleece, in good condition, on them. I reiterate my thorough conviction, that the Australian Merino, if well-bred, and the purer the better, will retain his fleece uninjured, as long as he lives, under the conditions above stated; and further, that any sheep containing the blood of the English coarse breeds will not retain his wool for more than two years, and often not for that period.

The most valuable wool to grow is a long fine "combing" wool; a wool that will comb "freely," that is, that will pass through the combs without leaving any "floss," or short portions of the fibre. Wool that "gnarls," or leaves any short wool on the combs, is called by the sorters "cloudy."

CHAPTER X.

ON YOLK IN WOOL.

THE influence of the yolk on the growth, condition, and value of wool, has, in my opinion, never received the attention it deserves ; and even in these days of improved, and, let us hope, improving, sheep-farming, very little importance seems to be attached to it. In the year 1838 or '39 I was engaged in selecting some young rams for a stud-flock, and had rejected a few very yolky sheep, when Mr. F. Bracher, one of the best and most successful breeders ever in New South Wales, came up and inquired why I had rejected some of them ; I answered him, "because they looked so black and dirty," and I shall never forget his reply. Having first closely examined them, he said, "Before you are grey-headed you will see the day when people will be as anxious to produce yolk as wool ; and when the salt-bush country is more opened up, the wool will be nearly value-less, in consequence of a scarcity of yolk." How these predictions have been verified, let the Riverine squatters say. It is a very general impression that heat produces yolk, and nothing is more common than for a shearer to say, "this is a fine warm day, it will make the yolk rise." This is entirely a misapprehension—heat never produces yolk, it destroys it—merely renders it into a fluid state, and causes it to run. Yolk in cold weather is in a solid or compact state, and when the temperature is over 80 deg. it becomes fluid, and in that state is freely evaporated by the sun. This is why we have so little yolk in our arid salt-bush countries. This excessive evaporation is always followed by more or less "tenderness" in the wool. In proof of these assertions, I adduce the well-known fact, that in cold or high countries sheep have the greatest amount of yolk— to wit, on the table-lands of New England, Bathurst, Monaro, and the upper portions of the Severn, McIntyre, Big River, Namoi, Macquarie, Lachlan, Murrumbidgee, and Murray. We find also sufficient yolk in the western districts of Victoria, and on the upper portions of her rivers, as the Avoca, Richardson, Loddon, etc. In Gipps Land too, yolk is abundant, in fact, the greater the altitude, the lower the temperature, and the more the yolk. On the Upper Goulburn, Ovens, and throughout the north-east district, this essential substance is plentiful. Remove sheep from any of these elevated regions to the lower parts of the rivers, and in the course of two or three years a portion, although not perhaps a very essential portion, of the yolk will disappear. The proper and indeed the only remedy for this consists in the selection of sheep for such localities possessing an abundant supply of yolk. Even in the highly favoured western districts of Victoria, I have seen wool on the back regions of the sheep become, after a very wet, cold winter, thin, light, and tender. This, doubtless, was caused by the heavy rains washing the yolk out, and the sun afterwards drying the wool rapidly, and evaporating a further portion, in some cases perhaps the whole of the remainder. Of course when this occurred, the seasons were exceptionally severe.

The consideration of this subject is of the last importance to sheep-owners occupying any of the salt-bush countries, the warmer portions of northern Victoria, Riverina, a great part of Queensland, and the north of South Australia. It is utterly impossible for any of these countries to produce a first-class wool without a copious supply of yolk. When there is an absence of this secretion, the wool becomes harsh, dry, and tender ; and on the back of the sheep, where the rays of the sun strike directly and continually, the fleece becomes "mushey"—not unlike a withered plant, dying for the want of its natural support—moisture. We cannot, as the sheep-owners do in England and Scotland, artificially supply a deficiency of yolk by "smearing ;" but we are told that our Merino sheep produce this substance more abundantly than does any other description of sheep. This I can easily believe, for all the English coarse-woolled sheep I have seen were perfectly white ; a colour which, in unwashed sheep, is highly objectionable, for reasons which are elsewhere assigned. The salt-bush and the other tracts I have named generate yolk very sparingly, and that little is soon destroyed by the fervid rays of a summer's sun. I do not say that we can entirely obviate this difficulty, but I think it may be considerably modified, by the use of the very same principles of selection which enable us to produce other desirable qualities in the fleece and carcase of the animal. In the first place, the rams introduced into the tracts of country referred to should be selected from flocks noted for their abundance of this secretion ; and your own ewes should also be chosen for their greater tendency to its production ; for the production of yolk is as essential to the sheep-owner as the production of wool, since, without the former, the latter will be all but valueless. In connection with this subject, it may be not out of place to urge upon the attention of sheep-owners in Riverina the very slight protection from the heat of the sun which many thousands of their sheep can get. Any person traversing the runs of Riverina, as I have often done, must have remarked that the wells, dams, and reservoirs, are usually formed at spots as remote from the timber as it was possible to place them; in some instances, from the peculiar features of the ground, this may have been unavoidable, but in the large majority they could, without any inconvenience, have been placed close to the edge of the timber. On these bare plains, near the watering-places, such an observer must often have seen large numbers of sheep—perhaps 4000 or 5000—split up into small mobs, and standing in circles, with their heads thrust under each other's bellies for shelter; clouds of dust continually rising from the stamping of their feet, a scorching hot wind blowing, and a vertical sun pouring its unmitigated heat upon the gasping creatures; and in this state and position of utter torture, they not unfrequently remain from nine o'clock in the morning until five o'clock in the evening. This would go near to burn a hole in a brick wall; what wonder then if it destroy the delicate fibre of a fine-woolled fleece. Surely, if sheep are worth keeping at all, they are worthy of a greater degree of care and attention than such a state of things betrays; and it is no imaginary or isolated case, but may be seen any summer day on the majority of stations in Riverina. Some may say that they have large paddocks without a tree within miles of them; if so, they may be assured that such portions of their runs are totally unfitted for summer use; and if they be not worth keeping for winter use alone, the sooner they are got rid of the better for the proprietor. But if there be no trees, is it impossible to plant around the watering-places, or at least, to sow seeds of, such of the more umbrageous kinds as are found in the district? Meanwhile, as a temporary mitigation of the evil, the man in charge of the well, or the boundary rider, might, for a few successive days, drive the sheep, after they have watered, into the timber, if it be not altogether too remote; after which the sheep themselves will seek the shade.

It is wools produced on runs of this description in Australia that the Cape and South America are, even now, successfully competing with; and all who have the interest of Australia at heart must devoutly wish that the foreign growers may ultimately drive the rubbish out of the market, and thus compel the wool-growers of this and the adjacent colonies to produce a better article. With our first-class wools, it is well known, neither the Cape nor South America can compete. Then why do we persist in an idle rivalry of inferior goods, when it lies with ourselves alone to produce an article which can defy competition? It must be particularly impressed upon the minds of sheep-owners that excessive heat destroys yolk, although in former days the direct converse was held to be correct, and sweating sheds were provided for the purpose of encouraging its secretion; an erroneous practice, which modern intelligence has subverted. Numbers of our sheep in the warm districts possess a fair amount of yolk after a cool winter, but this is again evaporated unless they be shorn early, before the sun has the power to destroy it. The early shearing recently adopted by many is, therefore, a step in the right direction; first, because there is less dust to injure the washed fleece, and next because there is more yolk to give the wool that peculiar softness so much admired by the wool-buyers and manufacturers. Even wool that has been hot-water washed, and all the yolk taken out of it in the process, still retains a part of its peculiar softness, and all its soundness or strength, provided only it has had a fair supply of yolk during its growth.

The Riverine, and indeed the whole of the salt-bush squatters, will have to improve their wool; and, by the pains and expense they have already bestowed upon the importation of rams from the western districts, I perceive that they have recognised the fact that the best and most effectual means they can adopt to achieve the desired end, is to employ rams purchased from the well-known breeders of that part of Victoria, for there the sheep possess in an eminent degree the qualities they require, viz., yolk, fineness, density, strength, and length of staple. A great object to be aimed at is, that all the fleeces should have a decided uniformity of character, and that character must be such as will ensure the highest money value. Now, the fleece to do this must be of the following description, viz.:—Long staple, say from three to three and a half inches, fine, close, soft and yolky, with none of the imperfections which I shall presently mention under their several technical names. I should be well satisfied with a ewe that yielded from 3lb. to 3½lb. of good, clean, well-washed wool, and reared a lamb at the same time. A ram should yield double the weight; and here I will observe that a great error was committed in the introduction and crossing with the Australian Merino of the English sheep. This was done with a view to increase the size of carcase, but its effect has been to render the wool of flocks so crossed of a nondescript character, and valueless; whilst the little measure of success attending it in the direction of size will soon die out, from causes which I have elsewhere indicated when treating of the Rambouillet.

E

CHAPTER XI.

WASHING.

So much has been written, and so many new plans invented, tried, abandoned or adopted, that it is quite impossible to say which is the best; but it may be said, without fear of contradiction, that no special system is suitable for all places. Any system whatever must be modified by the quantity of water you have at your disposal, its chemical peculiarities, its position, and a host of little particulars which will make themselves understood at washing time, although, perhaps, already familiar to experienced men. Without condemning, or, indeed, disparaging any of the new systems, I must say that I have not yet seen, at equal cost, any washing performed which left the wool in better marketable condition than did the simple plan of cold water and the spout. For more than twenty years, on one station, I adopted this plan, having a good running creek of clear, soft water, with a gravelly bottom. Half the day's washing we soaked in the evening, and the other half in the morning before we commenced to wash the first half. We did not spout-wash any that came clean out of the pool, but only such as still had black tips on the ends of the wool, or a dark shade about the rump or ribs. These, in passing up the race, were drafted into a small pen, washed with soft soap—which was kept in a bucket handy for the purpose—and afterwards spouted until they were perfectly clean. When the water is soft, and the sheep moderately clean, not more than 15 or 16 per cent. would require spouting. As spouting is generally practised, the clean, white-woolled sheep get as much as the black-tipped, close-woolled, dirty ones; this is to put the constitution of the animal to a test for which there is no necessity; for when a sheep is thoroughly clean, you cannot improve it by keeping it immersed in water. This system—other things being equal—I would still pursue; but in every case, the judgment and practical skill of the owner or manager will best direct them how to avail themselves of the natural facilities they may have at their disposal. After soaking, it is necessary to keep the sheep pretty close together, so that they assist in soaking each other. For this purpose, they should be put into a race of about one hurdle in width, and at every four or five yards a hurdle should be placed across the race, so as to form a series of small yards or pens; for, if left in too large masses, some of them might get smothered. Whilst in these pens, water, drawn from elevated tanks by means of a hose, should be scattered over them, so as to well moisten their backs when dried by the sun; this will be found of great service. They should not be allowed to lie down, for if they do, the new dung will communicate to the wool a green stain, which nothing less than soap and hot water will remove. Many persons, when soaking their sheep, use the crutch as freely as if they were washing them. The crutch should never be used in soaking, except just to dip the heads of the sheep, or any part that appears to be dry. All that the operation is intended to effect is their thorough soaking, so that the wool is saturated; but by crutching you deprive them of a considerable quantity of yolk, which is required to act as soap in the subsequent washing. For this reason, also, it is very objectionable to wash sheep a second time. Occasionally, owing to a break in the weather, or some mismanagement, it is deemed necessary to wash them again, the wool having become, in appearance at least, as dirty as it was before they were submitted to the operation at all.

It is impossible to wash sheep a second time with success, because the first washing dissolves and removes the saponaceous constituent of the yolk, leaving only a kind of gluten, to which every particle of dirt suspended in the water will adhere, and the fleece becomes full of them. This is termed by wool-brokers and classers "moity," or full of motes. Sheep, after being washed, should be kept four clear days, if in a country where this can be done, without their becoming dirty or tinged at the points of the wool; and, if practicable, they should be turned into a paddock, and not kept close folded together, lest they discolour the wool by rubbing against each other. It is essential, also, to keep them as cool as possible, so as to encourage the production of a fresh supply of yolk. It must never be forgotten that cold weather developes the yolk in a solid state, and heat renders it fluid, and disseminates it through the wool, whilst, if the heat becomes excessive, the yolk evaporates readily.

It has been asserted that when sheep, after soaking, remain long without being washed, the wool never exhibits a good white or bright colour, but assumes a kind of dingy appearance. This may possibly be the case after soaking in hot water, but I never found the cold soak to produce such an effect, even when the animals have remained wet for twelve hours after soaking. I am of opinion that hot water takes too much of the yolk out of the wool, causing it to handle harshly; and in this desiccated condition it has to remain for perhaps six or nine months, before it gets into the hands of the manufacturer. If, after washing in hot water, the sheep could remain eight or ten days, so as to permit the yolk to rise again, the plan might answer; but to keep them that length of time would render them as dirty as ever. Now, in the Riverine, or indeed in any salt-bush country, the usually friable nature of the soil causes it to pulverise readily; clouds of dust arise, and the sheep cannot be kept more than two or three days after washing, and even then they are often dirty. Hot water washing may be " the thing" in some districts where they have a close sward and little dust; but perhaps in the majority of districts —certainly in very many—it is a mistake, as the wool must be shorn before one particle of yolk can possibly rise, and in this dry and sapless condition it must be submitted to the keen eye and delicately appreciative touch of the European buyer, who, it is well known, prefers, other things being equal, the cold to the hot water washed article. There is no doubt that the yolk rises much more quickly after cold than after hot water washing; and the necessity for using hot water is by no means so pressing in the warm, salt-bush country as in (say) the western districts of Victoria, as, in the former, sheep secrete but little yolk, and very few of them get black tips to the wool. In fact, the salt-bush wools contain nothing in the way of dirt which soaking and a good spout will not take out.

CHAPTER XII.

SHEARING.

THERE is no branch of sheep management in Australia that requires a thorough and effective reform so much as Shearing; and no one who has either presided over the operations of a wool-shed, or filled any post of responsibility connected with the annual shearing, will deny that the system under which this important work is performed is beset with the gravest imperfections. The flockmaster invests a large capital in the purchase and improvement of sheep and run, spares no expense, no time, no care, no toil, in putting upon his flocks the best fleece they are capable of bearing, washes them with soap and hot water, by means of an expensive apparatus, and brings them into the shed in as perfect a condition as possible; and when there, a careless, hasty, or unskilful shearer so mangles and mutilates both sheep and wool, that a considerable portion of the squatter's labour and expense is thrown away. The remedy for this is a question surrounded with difficulties; and although we may for the present despair of a perfectly successful solution, it may not be amiss to suggest such palliatives as the writer's experience has dictated. To do this we must begin by recognising the causes of the evil; and of these, the first is the want of unanimity amongst the squatters themselves. If these gentlemen would only instruct their superintendents and overseers to insist upon the work being performed carefully, without hurry or bustle, and strictly in accordance with the directions given by an experienced and capable manager, the hands would soon recognise the fact that they could no longer be permitted to ruin their employer's property, and sully the reputation of his brand, in order that they (the shearers) might secure a big cheque. Again, a system of granting, to good and careful shearers, a written certificate, should be unanimously adopted by the sheep-owners, and men presenting these should be employed in preference to those without such testimonials. There would be no danger of your shed standing still for lack of hands; on the contrary, the pick of the shearers would gladly resort to stations so conducted, because there they would not be compelled, for the sake of a trifling numerical superiority over others, to turn their sheep out of hand in an unworkmanlike manner. Persons of limited experience, and, perchance, of not very extended sympathies, are apt to regard all shearers as a pack of necessary evils. This is a grave error, for it is not alone a false deduction from imperfect premisses, but by moulding the conduct of shed managers in accordance with a mere prejudice, it degrades the really conscientious and able man to a level with the "tomahawker," and eventually, perhaps, causes him to sacrifice his own principle and his employer's interest to the pressure of expediency. At the present day, the sole object in almost every shed would appear to be, to get the wool off the sheep in the shortest possible time, without much regard to how it is done, provided the animal be completely denuded of his covering. To see a set of shearers at work, one might suppose that their very existence was dependent upon the speed they should make with every sheep they catch. Now shearing is an operation requiring some skill, reasonable despatch, and, above all, great care; and the neglect of this last essential is the cause not only of great injury to the fleece, but often of death itself to the sheep, from the cutting and hacking which both sustain during the operation. The majority of shearers have either a bad method of performing their work, or they have none at all; and too often it occurs that the parties in charge of the shed are incompetent to correct their men. The proper method is, first, to clean all the points, the crutch, and the belly wool, and let this be swept aside, and taken away at once by the "picker-up." Then the neck should be carefully opened, not with the shears slanting, but by small, careful blows, until the wool is parted, and the shearer has secured a "good face" to his work. The greatest injury to the fleece takes place on the back, and is caused by the operator not raising his hand, so as to keep the points of the shears close down to the skin; this is known as "cutting through," and it takes place when the sheep is being shorn on one side; and in shearing over the back, the points of the shears cut nearly, or quite through the fleece, from the inside to the out; and when the animal is being shorn on the other, or "turning out" side, the shears are again pointed upwards, and the cuts on the first and last side overlap each other, forming a sort of vandyke line, and causing the fleece to part into two halves all along the back. Now this is altogether inexcusable, and I would not keep in the shed a man who shore in this way if he would pay me for permission to work. The matter is, however, very easily detected by spreading the suspected fleece on the table, with the cut side uppermost, when you will at once perceive where the shears have cut through. The fleece is then said to be "cut under and over."

No good shearer makes second cuts; the fact that wool has been left by the first cut proves that the shears have not been held properly, and the wool thus removed by a second cut is perfectly useless, entails a severe loss on the manufacturer, and greatly lowers the repute of the brand in the markets. The only place where a second cut can be tolerated is about the points, where the wool hangs in tangles, and, perhaps, cannot be cut through at the first blow. Many, in opening the belly, make a slanting cut, commencing some distance up the ribs, and thus leaving on the rib an inclined plane of wool, which has afterwards to be taken off, and is, of course, shorter than its proper length by just so much as was removed at the first cut; but the bellies should always be shorn first, and the wool removed, so that the classer may be spared the delay of separating it from the fleece. Some shearers may tell you that they learned to shear in this objectionable way, and cannot shear in any other; let such a man leave your shed at once, and learn the right method somewhere else.

The presence in your shed of a leading man, or, in shearers' vernacular, "a ringer," is most objectionable. This man is supposed to be privileged to call "Smoke, oh!" or "Knock off!" etc., and he assumes a sort of leadership which leaves the superintendent himself a divided authority only; and this is plainly subversive of that discipline and good order which should supplant the bustle and haste before deprecated.

As one consequence of their hurry, many shearers take no trouble to keep their fleece entire, but trample carelessly upon it, and allow the sheep's legs to get entangled in the wool, and, should the sheep kick and struggle, the fleece is broken, literally, into "pieces." This must be checked at once, and peremptorily.

Many men who call themselves shearers can never be taught to hold a sheep properly, and the animal, in consequence, is struggling violently all the time of being shorn. Now, this forms, perhaps, the most certain test of superiority amongst shearers; and you may safely dismiss from your shed any man who is constantly fighting with his sheep. Besides the injury done by such a man to your wool, he will frequently lose temper in his conflicts with the refractory animal, and inflict great torture, perhaps even fatal injury, upon the creature, by compressing its nostrils, violently wrenching its neck, kicking it, or forcibly pressing his knee upon its ribs and flank. Cutting through, however, and a habit of continually making second cuts, are the most objectionable characteristics of the unskilful shearer, and these should be especially noted. A sheep may be shorn so close as to satisfy the most exacting employer, yet it may be shorn very badly; and the only conclusive test of good shearing, or the reverse, is to be found, not on the outside of the animal, but on the inside of the fleece. I would not require the sheep to be shaved, as they term it, provided the fleece be taken off moderately close, perfectly even, without second cuttings, and without breaking it.

To manage a large shed is no sinecure, and demands, on the part of the manager, a combination of patience, skill, firmness, and tact, not easily found united in one person. It is an unwise practice to bounce or find fault with a man in the presence of his fellows, especially whilst he has a pair of shears in his hand. The proper time to remonstrate with a man is at night, or at dinner time; then, if you are dissatisfied, tell him so, and make him clearly understand why. Inform him that you make it a rule never to speak to a man more than once, and that if he cannot so alter his hand as to give you satisfaction, he had better say so, and leave at once. If his work be not improved after this hint, pay him off, and let him go, but do not go on ruddling badly-shorn sheep until serious loss is incurred.

I think it very bad policy for an employer to stop or deduct from a man's wages on account of his work not satisfying. If he does not suit you, it is your business to find it out before mischief or loss has accrued; tell him so decisively and quietly; pay him his wage, and let him go. A regulated and equable kindness will always do more than coercion, howsoever applied; and when a manager or proprietor condescends to scold or bandy words with his labourers, he forgets and abuses his own position, and his men will soon follow his example, probably with much more vigour and effect.

There is not, whether on farms or stations, any kind of work performed in so hurried and careless a way as shearing; yet, in the entire routine of either, there is none, in this country, so financially important. Now, to alter this system, is the very groundwork for reform; and it can be easily done. Just mark what these hungry, bustling men tally; confine them to ten or a dozen per day less, and do not allow them to leave the shed until the other shearers go. In conclusion, I must remark that from 60 to 80 wethers is a good fair day's work for a good and careful shearer; and in the whole course of my experience, I never met with a dozen men who were able to shear 100 per diem to my satisfaction.

In many parts of Victoria and New South Wales, especially in the Riverine country, great difficulty is experienced in keeping the sheep clean after washing, and getting them from the paddock to the wool-shed without dust. It is my conviction that the greater part of the dust accumulates in the fleece when the sheep are being cut off in the paddock, and in taking them from the paddock to the wool-shed. To obviate this, an ample paddock should be fenced off around the wool-shed, and never used except at shearing. The shorn sheep should not be turned into it, and no more of the unshorn sheep than will serve for each day's work.

I cannot conclude without a few remarks on the shearing of lambs. The practice, hitherto, has been to do this at three or four months old. I think this a grave mistake. In five years out of six it could be avoided with advantage; in the March or April lambings it may be necessary, but in the May and June lambings (and these are the majority) it is nearly always injudicious. The wool from lambs of three or four months old rarely pays more than the expense of shearing, sending to market, etc., and the operation completely destroys the ensuing hogget-fleece. I admit, nevertheless, that there are seasons when it becomes necessary to shear the lambs at this early period of their existence; for example, after a heavy fall of rain in the winter, when there is likely to be a flush of grass, and therefore of grass-seed. With this exception, however, if the lambs were left unshorn until the commencement of the following season, they might be shorn, say in August or early in September, when they would have an average of about 14 or 15 months' wool on them, which would be of the most valuable description. This arrangement would assuredly pay the flock-masters. In the northern parts of Victoria, and throughout Riverina, several squatters have, on my recommendation, adopted this system, and they are quite satisfied with the results.

CHAPTER XIII.

ON FLUKE, AND THE GRASSES FAVOURABLE TO ITS DEVELOPMENT.

THIS disease is prevalent in certain localities of all the Australian colonies. In England, it is almost exclusively confined to low, swampy, or marshy ground; in Australia, it is rife and virulent only on the high lands. It is principally confined to the heads of those rivers which take their rise immediately from the main or dividing range; on the whole of the western flank it is the most malignant, and the nearer the head of the range, the more fatal it becomes. Commencing, for example, to trace its course from the head of the Condamine (the most northerly river running westerly in Queensland), we pass down by the Severn and its tributary creeks, all of which are Murray waters, until we reach a distance of 60 or 70 miles from the foot of the range, when we approach the low, level land. Here, simultaneously, the fluke disappears and saline vegetation commences, continuing to prevail in greater degrees, until the true salt-bush country is reached; and I am disposed to attach great importance to the coincidence, because precisely similar phenomena characterise many similar tracts of country with which I am well acquainted—to wit, that through which the McIntyre flows, the Big River, the Namoi, the Peel, the upper portion of the Lachlan, the Macquarie, the Murrumbidgee, the Murray, and the Goulburn. Even the hard, dry, slatey, and trappean formation at the head of the Turon is harassed with fluke, and it penetrates to the westward of Bathurst as far as Molong. In a word, you may find fluke on most of the high lands, but as the rivers descend westward to the salt-bush country, fluke ceases to be known (Youatt, p, 458). Now, although the high lands produce a succulent if somewhat sour grass, yet the varieties of grasses are very limited, and the saline vegetation entirely wanting. The inference is one which I drew years ago, and have seen no reason to modify, namely, that fluke is due to continued monotony of food, together with the deprivation of saline herbage. Our experience is now sufficiently extended to entitle us to draw conclusions and inferences; and, amongst the former, we may state it is quite clear that our low lands, however soft and marshy, provided they bear salt-bush, or even a sufficient variety of grasses, will not generate fluke. If, in this hemisphere, feeding upon land that has been flooded would affect sheep with either fluke or rot, the sheep on the banks of the Condamine, and many other rivers, could not possibly escape. Every successive year that country is flooded, the water remaining on it sometimes for a month or more, yet no sooner does it subside sufficiently than the sheep go eagerly on to it to feed, and without any ill effect whatever, except foot-rot, that I could ever hear of. The same may be said of the Darling, where, at times, there are hundreds of square miles covered with water at least six or eight inches deep, no means of drainage, except absorption and evaporation, the thermometer at no less than 140° or 150°, and the ground so soft that, at every step the sheep takes, his hoofs sink three or four inches into the soil.

According to the English writers on the subject, this state of things would ruin every sheep put upon the land. That sheep may contract fluke on flat lands, I do not doubt, but it can only be at certain altitudes, at least in this country. During twenty-six years of practical experience amongst flukey sheep, I have detected it in sheep of all ages, even in lambs of no more than three months old. I once considered the disease hereditary, but relinquished that idea after carefully examining, without detecting any symptom of the disease, the livers of a great number of lambs which died at ages varying from two to fourteen days. That it shortens the lives of sheep I have no doubt; as, in the district to which I more especially refer, few sheep attained to a greater age than about seven years; indeed, after their fourth year, the ewes began to get thin and puny, and but few of them reared lambs. If the disease be not too far advanced, they may be cured by sending them to a salt-bush run; this I have frequently tried, with very satisfactory results. But if cavities, and bony or calcareous knots have already been formed in the substance of the liver, the organ can no longer properly perform its functions; and, unless in some instances of young and vigorous constitutioned animals, the sheep is incurable. This, also, I have experimentally proved by, on several occasions, removing such sheep from Bathurst to Wellington, or the Lower Lachlan. How the fluke gets into the liver, whether, as some have conjectured, it is, under certain hitherto unrecognised conditions, generated there, or whether it is taken into the stomach with the food, I am by no means physiologist enough to determine. In England, fluke and rot are almost convertible terms; but in this country there must be points in

which they differ widely; for here I have never known young
sheep to die of fluke, except in a few very doubtful cases;
whilst rot in Europe is fatal to both young and old. Some
seasons are far more favourable to the development of fluke
than others, nor is there any doubt as to the character of those
seasons; the old sheep that are infected die much more quickly
in a wet, cold winter, than in a warm, dry one.

It has often struck me as worthy of note, that in a flock of
the same age, bred from the one flock of ewes, and always
running together from the time they were weaned, some sheep
would escape the fluke altogether, and when killed their livers
would be as clean and free from fluke as those of any sheep in
the world, whilst others would be severely affected by the
disease. The sound sheep, too, were generally the best conditioned. It is very rarely that a flukey sheep will retain his
condition, after he is four years old; and flukey ewes especially
get "bottled" in the neck, and die off rapidly at five or six
years of age. We have indeed much to learn about fluke and
its cause or causes—its effects we know only too well. In
gum, stringy-bark, and granitic countries, where the grass is
long, green, succulent, and sour, fluke may be pretty confidently expected to prevail. I have often heard it remarked
that those among the squatters who annually send their sheep
from Riverina up the mountains, will one day regret it, as their
sheep will become flukey. There is, in my opinion, no danger
on that score, so long as the flocks are brought back to winter
on the salt-bush. I do not remember ever to have seen fluke
on the eastern side of the dividing range, but a disease prevails
there which is nearly as bad. About the Clarence and Richmond rivers it is called "gut-rot;" at Western Port, near
Melbourne, and all along the coast to Port Gawler, in South
Australia, they call it the "coast-disease;" but in many places
it extends from the coast 100 miles inland. It is a kind of
chronic diarrhœa, for which neither a cause nor a cure has yet
been discovered. It is a frightful disease, and in some seasons
terriby destructive; its principal feature being a continual and
violent purging, under which the sheep wear away to mere
skeletons. The only palliative known is the same as that
applied to fluke, namely, instant removal to the salt-bush
country; in fact, the salt-bush appears to present the allevia-
tion or the cure for all the diseases incidental to sheep in Australia.

For many years it has been known, that on the eastern
waters there is very little country good for sheep. The best
zone—if I may be allowed the term—for both the growth of

wool and the health of the sheep, is intermediate between the western slope of the main range and the commencement of the salt-bush country, where the grasses are intermixed with a proportion of saline plants, sufficient to satisfy the instinctive desire of the animal for that condiment, or therapeutic, and to preclude the attack of fluke. It is unquestionable that wherever there is no saline herbage, and the grasses are restricted to very few varieties, in that place fluke is imminent. About the Fiery Creek country, Streatham, and the district around Geelong; in fact, in many other parts of Victoria, such as Mount Emu, Mount Elephant, etc., the sheep are free from flukey tendencies, although those localities are destitute of saline herbage; but its place is effectively supplied by an unusually varied assortment of the best grasses; and salt does not appear to be required. It is quite easy to discover when your sheep require salt; and if there be any dry creeks on the run, and the ground is in the least impregnated with salt or saltpetre, you will find the banks partially undermined by the sheep licking and gnawing the saline earth. When this is observed, no time should be lost in giving them salt, mingled with one-eighth part of pulverized sulphate of iron. Most people give rock-salt, which they place in small heaps at various parts of the run, where the heavy dews of our climate or a few showers of rain dissolve and waste much the greater portion of it. The method of administering the salt which I have invariably practised, and can therefore recommend, is to mark off a piece of land, level ground, say about twenty yards square, and sow the common salt on it as if you were sowing grain, in the proportion of 1oz. to each sheep. If the sheep have been long in need of it, they will appear almost crazy with delight on first tasting the salt. Some of their antics are extremely droll. For instance, if you hold a white handkerchief high up, even over your head, they will jump at it, mistaking it for salt, and endeavour to reach it by rearing on their hind-legs, and, with the utmost boldness, resting their fore-legs on any convenient part of your person. I speak not of pet sheep, but of the members of an ordinary flock, bush-fed. The necessity must indeed be urgent which could so entirely overcome the timid nature of the creature, and the merest common humanity should concur with prudence in anticipating the want. A week after the first dose, another ounce may be given, and repeated weekly until you observe that the sheep have become careless of it. These salted sites will form licking-places, which the overseer should occasionally visit, and should he find a number of sheep about, or detect their teeth-marks on the ground, he may be assured that they require more salt, and not an hour should be lost in supplying it.

Many sheep-owners in the district where I acquired the greater part of my experience in fluke, either from ignorance, carelessness, or, worse still, from parsimony, neglected to give their breeding ewes salt, and the consequences were so frightfully disgusting to witness, or even to think of, that I hesitate to record them, and would certainly decline the revolting task were it not for the really solemn lesson they convey. To the educated, humane, and intelligent squatter of the present day, the narration must appear almost incredible; yet, in the neighbourhood of Bathurst, King's Plains, Mount Macquarie, and on the south-east of the Conoblas, there are still alive, I dare say, hundreds of people who could testify to the rigid accuracy of my statement. At the period I speak of, sheep, especially ewes, were shepherded in flocks of from 400 to 600 sheep each, and, during lambing, there would be three men to a flock, viz., one hut-keeping, one with the ewes and lambs, and one with the lambing ewes; yet, in spite of the watchfulness of the men, I have seen ewes, not once, but hundreds of times, after they had dropped their lamb, turn round and gnaw off its feet and tail, and then greedily devour their own "cleanings." It was quite a common thing to see half-a-dozen or more ewes running after one that was about to lamb, until she lay down. I have seen the poor animal rise, perhaps half-a-dozen times, and endeavour to repel her assailants. They, however, would not be repulsed, until the lamb was dropped, and no sooner was this effected than her persecutors—who, by the way, were termed "midwives"—would rush in, seize the lamb, and rend it into fragments, like hungry wolves. I have also seen them eat a portion of the "bearing" of the yeaning ewe; and all this took place simply and solely for the want of a little salt. The reader will say—Why not take these "midwives" out of the lambing flock? Simply because it was useless to do so; for no sooner was one lot got rid of than another drew out and performed the horrible office. Whilst giving salt to a flock of sheep that have been a long time without it, and are eager for it, you may walk through and through, and even handle them as you like. They take no notice whatever of men or dogs, and no number of either could disperse them whilst any salt remained. By adopting the plan I have sketched out, sheep need never be reduced to this terrible condition; but the salt grounds must be vigilantly watched, and, at the very first symptom of an inclination to gnaw the ground, a fresh issue of the salt mixture should be supplied.

Many persons assert that my plan of administering salt is an extravagant one, and that to give it in long wooden troughs is best and most economical. But the trough question is open to objections, which, I submit, are fatal to its adoption, at least in the first instance. For example, if the sheep were in extreme need of the salt, and eager to obtain it, they would smother one another in their struggles to get at it, and the stronger animals would monopolise the whole; and, getting thus much more than the proper dose, they would soon suffer extreme thirst, and gorge themselves with water until rupture of the intestines proved fatal to them. This I have seen occur. When the longing for salt has abated, the troughs might be used, if the manager thought fit. Other parties have advocated the trough system because they thought the animal would gnaw the ground; but sheep do not gnaw the ground unless they require salt, and whilst the article is to be found on the surface, the tongue alone, and never the teeth, is used in securing it. The space of ground prepared and used for the purpose must, of course, be in proportion to the number of sheep to be treated, so that all the individuals of the flock, both weak and strong, may be certain of getting a share of the medicine.

I observe that many of the Victorian sheep-owners are sending their flocks in large numbers to summer at Omeo Plains and other parts of the mountain country. This system may do very well for those who have salt-bush country to bring the flocks back to for lambing and shearing; but to those who have neither salt-bush, saline herbage, or variety of grasses of any kind, it may prove a perilous experiment. Finally, it should be observed that the best period for administering the salt is after a good rain has left abundance of water near the salt-licks, as the sheep, especially at first, will be very thirsty.

CHAPTER XIV.

FOOT-ROT.

THIS very annoying disease has been much modified, both in extent and severity, since the fencing of runs, and their subdivision into paddocks, became general. The sheep, freed from the control of the shepherd, are now at liberty to exercise that almost unerring instinct for which, under the old *régime*, the limited reasoning powers and doubtful discretion of their custodian were substituted. In wet weather, for example, sheep, if left to themselves, will invariably feed and camp on the highest ground; or, should there be no high land within the paddock, they will be found on those spots which are driest, especially on sand. Most of the British writers on sheep husbandry are disposed to consider Foot-rot contagious. Never having visited Europe, it is not for me to controvert the doctrine; but in New South Wales, Victoria, Queensland, and South Australia, my experience fully bears me out in asserting that such is not the case. The gentleman for whom, during many years, I acted as superintendent, held stations in Victoria where Foot-rot abounded, and other dry, salt-bush runs where Foot-rot never occurred; and it was part of my duty, during the winter months, when Foot-rot most rages, to inspect carefully all the diseased sheep on the country, and cull out the infected, for removal to the drier runs. This I did, year after year, and the diseased sheep, after being properly pared, but without the application of any dressing, were turned into the flocks on those dry runs. Yet I never saw a case of infection; on the contrary, if the weather kept pretty dry, the sheep got well and sound as soon as the hoof had had time to grow. The disease cannot be so virulent here as in England; or there are conditions attending it there—perhaps climatic, perhaps pathological—from which we are exempt. What, in this country, we term Foot-rot, is merely an overgrowth of the hoof, and this is encouraged by a rich soil, formed of decomposed vegetable matter, containing no sand or gravel, and therefore affording no sufficient attrition to keep down the growth of the hoof. In fact, during wet weather, the hoof grows faster than the sheep can wear it away. The crust, or wall of the hooves—as well the outer as the inner wall—then bends in over the sole, foreign matter is collected, and the sheep becomes lame. Colonial Foot-rot often appears at the top of the foot where hair meets hoof, and a fungous excrescence forms around the coronet. This arises from the unusually rapid growth of the walls causing all the stress of locomotion, and the sheep's weight, to be borne by the walls alone, without the aid of the sole; when the upper part, already weakened and thinned, detaches itself from the flesh around the coronet, and the cup, or flange, thus formed, becoming filled with foreign matter, irritation, amounting to low chronic inflammation, ensues, and unhealthy granulations appear. This is the worst and most obstinate description of Foot-rot, but, fortunately, it is less prevalent than the other affections arising from the like causes. A sheep, whose hoof has parted in this way from the coronet, should, if in condition, be slaughtered for rations, as it is sure to lose the hoof, and, perhaps, its fleece, in consequence of the symptomatic fever accompanying the disease. Another kind of Colonial Foot-rot is called a " Scald," and is usually unaccompanied by any enlargement of the hoof. This is often seen to occur after a long drought, when the rain has softened the foot, and excoriation follows and may be produced by the friction of long grass accumulating in the cleft of the hoof after the rain sets in. It is easily cured, by immersing the foot in a weak solution of arsenic, or a strong one of lime-water.

In the treatment of such sheep as have the walls of the hoof folded over the sole, I first cut away all the superfluous hoof down to the level of the sole, giving the foot as nearly as possible the natural shape; I then cut the toes short, and bleed by pricking the toe with a strong needle. If preferred, you may bleed low down the thigh or forearm, but as near as possible to the seat of the mischief. The sheep should afterwards be placed on short grass, where the foot can be kept as dry as possible. I never applied any dressing, except in cases where the hoof had parted from the coronet, and then only a strong caustic; or, better still, the actual cautery, to destroy the fungus. The latter plan I have found very serviceable, as it substitutes the healthy wound of a burn, in place of the malignant ulcer of the disease. It is commonly considered that the mere richness of the soil has great influence in developing Foot-rot. In as far as rich soil retains moisture in it, or holds water on its surface, for a longer period than poor soil, it may have this effect, but certainly no farther. I have invariably found soils that retained moisture longest most productive of this disease. In hilly, stony countries, sheep are not subject

F

to this disease, simply because water cannot lie in such situations; and the stone or gravel affords the indispensable amount of attrition to the hoof of the animal. The fact of the ground being hilly would, however, not be sufficient of itself to avert Foot-rot; it must also present the other condition—namely, sufficient grit, gravel, or stone, to wear the hoof down as fast as it grows. By way of illustration, we may mention Bryan's Creek, Muntham, and many other stations in that neighbourhood, all of which are sufficiently hilly; but the soil contains no element to afford attrition. Whether the soil, then, be rich or poor, matters not a button; whether the soil be wet or dry, is the main question; and, subsidiary to it, whether there be sufficient grit, gravel, or stone. Darling Downs, I am credibly informed, during the first 15 or 17 years of its occupation, was quite free from Foot-rot, although within the tropics, and subject to periodical inundation, during which the whole of the low-lying lands were completely submerged, often for six or eight consecutive weeks, yet the soil here is of the very richest description, and literally stoneless. Many of the Darling Downs squatters have high, stony ranges at the backs of their runs, where, in former times, they sent their sheep until the waters had subsided, and the ground dried sufficiently to admit of their flocks being brought back to the lower country. But squatters there have been no wiser in their generation than squatters elsewhere; they have taken to over-stocking, and left no grass on the ranges to meet the annual emergencies, and they are, therefore, compelled to remove their sheep from the high to the low lands, before the inundation has subsided. The sheep, having been half starved whilst yet on the ranges, take the water like spaniels, and remain in it all day picking at the tussocks, or tufts, on the higher patches, which have their heads just out of water. In New South Wales, this is called "bogeying for grass." This constant immersion of the hoofs stimulates their growth, at the same time they are subject to no attrition, and I am told that Darling Downs can now vie with any station in Victoria for Colonial Foot-rot.

CHAPTER XV.

SCAB AND ITS CURE.

THE various modes of dealing with this terrible pest have long been well known, but the rapidity with which it appears to be spreading leads me to apprehend that not yet has it received, in Victoria, the degree or the kind of attention which can even partially subdue it. All the adjacent colonies have succeeded in cleaning their sheep; in New South Wales and Queensland the disease is eradicated, and South Australia also is perfectly free from scab. Victoria alone, with all her wealth, science, and energy (must I not add—all her quackery, too?), has not only failed in subduing the foe, as too plainly appears, but has permitted him to gain a palpable and very decided advantage. Wealth almost unbounded (and certainly always most liberally applied), the first mechanical ingenuity; energy and assiduity of no common order—all have been employed in combating this most ruinous malady, and all with less than no avail.

It appears to me, then, that a solution of the simple question—How is this? can be sought in one direction alone. The primary error of the squatters has consisted in patronising the quackery to which I have alluded; they have suffered themselves to be prevailed on, by interested persons, to use some one of the many worthless compounds designated "scab specifics." Doubtless the gullibility of the squatters, as well as the cupidity of their advisers, is in fault, but the Scab Act heretofore in force was so loose and indefinite in its provisions, that it almost offered a premium for its evasion or infringement; and it is to be hoped that the new Act will contain a clause making it imperative for the sheep-owner in future to use as a detergent the well-known and perfectly efficacious remedies, tobacco and sulphur combined in certain proportions according to circumstances, and mingled with water heated to 110 deg. or 112 deg. Fahr. This mixture is simple in its application sure in its effect, cheap, and perfectly innocuous, whether to the health or the fleece of the animal, if applied at the proper time—viz., after shearing. If, however, the use of trashy specifics cannot be wholly forbidden, let the penalty, in case of failure (which is certain), be doubled. Further, it should be made compulsory on every one dipping his sheep to furnish to the inspector of the district, for transmission to the Chief Inspector, a statement of what ingredients were employed. From records thus compiled the Chief Inspector, after, say, six or twelve months, could at once point out which ingredients had cured the greater number of sheep. This suggestion is no novelty; it was put in practice whilst scab was rife in New South Wales, and that record is considered to this hour (and never more so than at present) a most reliable and valuable fund of information; such, indeed, as the diseased condition of Victorian sheep might but too soon call into use.

A subsidiary cause of failure, in attempting to cure scab, is to be found in the difficulty of exhaustively mustering paddocks, so that not a single sheep shall escape treatment. Some few sheep are very commonly overlooked, especially on thickly timbered, scrubby, or broken ground; and should foot-rot be in the flocks this difficulty is greatly increased.

Every pains should be taken to collect all the sheep, for if this be not scrupulously done, the dipping is utterly worthless, and of no avail. There is one clause in the new Victorian Bill which, I submit, merits especial commendation; indeed, the measure would prove utterly valueless without it. I allude to that clause which enacts that the finding but one scabby sheep in a flock shall be sufficient proof that the whole flock is scabby within the meaning of the Act. This is as it should be, for whether the flock consists of only one hundred shepherded sheep, or of twenty thousand at large in a paddock or inclosure, it would be quite impossible to take one scabby sheep out of either lot, and be prepared to declare the remainder clean. This clause, and that which empowers inspectors to levy fines, and enables them also to enforce and recover them, form undoubtedly the gist of the Bill; indeed, so wise and comprehensive do they render the measure, that I regard them, if honestly and strictly enforced, as forming of themselves no bad Scab Act. That clause, also, which provides that all sheep coming into Victoria from any adjacent colony shall be examined by an inspector, is an excellent one. It possesses a prospective efficacy, which may hereafter prove to be most valuable.

To proceed to practice. Say you have a paddock grazing 6000 scabby sheep, all of which have been completely mustered, not one left at large. Commence on the first day by dipping 2000, *correctly counted,* and after dipping, as they leave the draining yard, put a conspicuous raddle mark or brand upon them, and hold them in hand, carefully shepherded, for 10 days;

proceed similarly with the rest, until all have undergone a first dipping. On the 11th or 12th day commence to repeat the process, with all the same precautions as were used in the first instance ; and, having put on them a second conspicuous mark to indicate that they have undergone the second immersion, ascertain that no other sheep have either got admittance to the paddock or joined the dressed flocks, and return the latter to their pasture. The ingredients I would use are those which have been commonly used for years past by the most practical men in the colonies (*i.e.*, by those who alone have succeeded in cleaning their sheep), and consist of from four to five ounces (according to quality) of good, sound tobacco, giving the preference to the unmanufactured American leaf, and three or four ounces of sulphur, to one gallon of water. The tobacco should not be boiled, as by that process the essential oil undergoes certain chemical changes which deprive it of its curative properties. The leaf should be placed in the water just as tea should be made, namely, whilst the water is boiling ; the vessel containing the infusion should be closely covered over to prevent the escape of steam. When dipping, if any of the sheep are very bad with the disease, and large patches of dry, crusted scab appear on their bodies, the particular animals so characterised should be placed in a pen kept solely for their reception, the rough, hard masses of scab removed, and the denuded parts spotted or dressed with tobacco water of double the ordinary strength, viz., eight ounces of tobacco to the gallon of water. The sheep so treated should not be dipped on the same day, or the weaker solution of the dip would wash the stronger one out before it had time to effect its purpose. If the sheep have been shorn very close, I would use, in the ordinary dip, not less than five ounces of tobacco to the gallon, because the pelt, almost denuded of wool, cannot retain the dressing ; and, for a similar reason, I would keep them in the dip from one and a half to two minutes, if the dipping take place within, say, 14 days after shearing. Much of this, however, must be left to the discretion and judgment of the manager, who should be able to tell when the newly-shorn wool has made such a start as will enable it sufficiently to retain the dressing, And here I may caution the manager against having his dip too deep ; from three feet to three feet six inches is enough, because the sulphur sinks, after it is once wet, to the bottom, and there becomes densely impact, so that unless it be stirred up by the feet of the sheep, it cannot lodge among the short wool as it should do. When the dip is used, and left standing for some time, the mixture be-

comes impregnated with a noxious gas, which may prove fatal
to the life of the sheep. To guard against this, the dip, before
being again used, should be thoroughly well stirred up from
the bottom, and these gases will evaporate. Always, too,
before commencing the day's dipping, stir the mixture well, so
that every particle of sulphur may be held in suspension by the
water. The sulphur acts upon the sheep merely as an anti-
contagionist, and not as a detergent, and when used in a dry
state it will not destroy the acarus. I have placed a number
of them on a piece of dry paper, and sprinkled sulphur over
them, and there they remained, during a period of nine days,
without any apparent diminution of their vitality. I have
also seen the acarus live after remaining three weeks on a
piece of paper that was kept moist by placing a wet cloth
under the paper three or four times a day. On the other hand,
I have seen the acarus perish in the course of four days during
the hot weather of summer; indeed, it is a proven fact that
the acari thrive best, and are much more lively in cold, wet
weather; whilst, in hot, dry weather, if exposed to the sun,
they die and dry up into mere dust in the course of a few days.
They will also bear the most severe frost without injury. The
tobacco, however, does not inflict sudden or immediate death
upon the acarus; care must, therefore, be taken to saturate the
entire frame of the sheep, including the head, lest the insect
there take refuge when dislodged from adjacent parts.

When a flock of sheep are strongly suspected of being scabby,
yet no tangible proof of the fact can be found, I have heard
people say, "Wait until the young feed shoots up, and it will
throw it out of them." This is a palpable and grave error,
for from it we might infer that scab is a disease of the blood,
or a constitutional malady; but, on the contrary, most people
are now convinced that it is a cutaneous or skin disease only.
It is not the young green feed which makes the scab appear
so suddenly at times; it is that which urges the young grass
into rapid growth—*it is the rain* that is, as I have already
shown, so favourable to the propagation of the acarus. The
rain and the young feed come almost simultaneously in our
fecund climate, and so the effect is mistaken for the cause. In
confirmation of this, I may state that I have seen a flock of
scabby sheep accidentally boxed with a clean one, in very hot,
dry weather, and not a sheep of the clean ones took the infection. In similar weather I have frequently seen one or two
scabby sheep get amongst clean ones, without any evil consequences ensuing. It remains for me at present only to add
that I have myself, in a very cold, wet district of Victoria,

treated many thousands of scabbed sheep in precisely the way I now recommend; that those sheep were immediately removed to clean runs in various parts of the country; and in no single instance has the disease re-appeared.

I may add a few words respecting the new curative proposed by Mr. Rowe. From my own personal knowledge, neither lime nor sulphur will destroy the acarus; how they may act when combined, I am not chemist enough to say; I, however, think it worthy of a trial. The standing of the gentlemen recommending it is a strong guarantee of its efficiency.

CHAPTER XVI.

DEFECTIVE WOOLS.

Stripey, or, as it is sometimes called, watery wool, once pointed out, cannot be mistaken. The parts of the fleece most commonly liable to it are the lower part of the fore-arm, the top of the wither, extending often along the back to the loin, also along the lower portions of the ribs, the belly, and a considerable distance up the flanks. It is not peculiar to any particular part of the country, or to the climate, but it is evidently perpetuated by breeding from stripey stock—ewes or rams—and as it is very frequently overlooked by sheep-owners and managers, it is of very frequent occurrence. When engaged in classing, I have not seldom had occasion to point out, to the owner or superintendent, the prevalence of this defect; when these gentlemen would reply, "We never saw our sheep with wool like that before; we think it must be owing to the late bad season, the wool being generally out of condition." This is an error; the fact is, it has been, perhaps for some seasons, becoming more prevalent, and therefore more readily discernible, until at length it can no longer be over-looked; and this is the natural result of continuing to breed from animals exhibiting this defect. In young sheep "stripe" first shows itself on the forearm, just below the shoulder, where the best wool on the whole animal should be found. If a staple be taken out at that place, it will be found entirely devoid of those natural curves which form one invariable characteristic of a really good wool—straight, and with marked indentations, about a quarter of an inch from each other, as though it had been passed through a crimping machine; indeed, it strongly resembles the " wavey " appearance sometimes given by artificial means to a lady's hair. When compressed in the hand, it will not spring up again—it seems lifeless, and has no vital elasticity, and handles more like flax than wool. By gentlemen of eminence in our wool-markets I am told that such wool is fit only for the manufacture of inferior goods, blankets, etc. To detect stripe in very young lambs requires much practice, a quick eye, and rapid discrimination; and should the lambs have been shorn in the same year they were dropped, they must of course be left unclassed until the following season. It is a fault which gets worse every successive year, and that degree of it which, on the lamb of a few months, or even of twelve months old, appears only slightly objectionable, at three years will pervade the whole of the parts I have mentioned, and the fleece will be comparatively worthless. In this advanced stage, the whole of the parts affected present a curly appearance, the pile is very thin on the pelt, and hangs down when it is long enough to do so. This defect I believe we have acquired from the coarse, long-woolled sheep of Great Britain, for I never noticed it before such animals were imported. It is almost needless to add, that such sheep should not be bred from, nor even retained in the flocks.

Toppiness is a defect in wool not perhaps of vital importance, but still not to be overlooked. It consists in wool having a toppy or pointed appearance, in consequence of the various fibres differing in length. On examining a staple drawn from a fleece having this defect, you will observe that at the bottom, or roots, of the wool you have a much greater bulk than at the top; whilst, in a perfect fleece, the two extremities should be of equal bulk. Toppy wools are usually of the long-staple description; and when the top is free, that is, has not a twist or gnarl, at the extreme end, the fault is not considered very material, except that it involves a loss of weight. When, however, the wool curls at the end, so as to form a quantity of noils, much of the wool is wasted in the manufacture, as these are merely dead wool, and break off on the least strain. In hogget wool—that is, the wool of sheep that have attained the age of twelve or fifteen months without being shorn—this defect is more excusable, and more common than in the fleeces of sheep more advanced in age, because in the younger animal all the fibres are said to be more pointed, so as to enable them to perforate the skin when the first fleece is grown. It is also surmised that the fibres do not all come through the skin simultaneously, consequently some of the fibres may be ten, fourteen, or more, days older than others, and of course longer. I am inclined to concur in this view, because after the sheep has been once shorn, and the fibres are cut to one length, toppiness decreases. I am not physiologist enough to account for the fact, but it is certain that lamb's wool is almost always more or less toppy.

Broad-topped Wool.—This is, unfortunately, a very common and most serious defect, and one which, I fear, is gaining ground, in consequence of unskilled persons mistaking it for density. Broad-topped wool, when seen on the sheep's back, presents to the inexperienced eye, in many cases, a well-grown, even-topped fleece ; but, upon closely inspecting and handling it, you find the tops of the wool interlaced, not exactly " felted," but something very like it. The whole fleece, when shorn, will, perhaps, hang well together by the bottom wool, whilst the tops appear in locks, from one to three or four inches wide or long superficially. Unlike toppy wool, however, you will find that you have the greatest bulk of wool at the top, which has a hard, matted feel ; and on attempting to draw a small staple out of it, it will cling and break at the top, in place of coming freely away. The matting, or interlacing, arises from the wool splitting from the top downward, perhaps, from half an inch to an inch ; not having then sufficient strength or rigidity to maintain its naturally vertical position, each filament of the split fibre turns down and becomes interlaced with the top portion of the fleece. The whole mass at the top then loses the strength and toughness of vitality, becomes, in fact, dead, and breaks with the slightest strain, causing immense loss and annoyance to the manufacturer. Amongst defective wools, there is none of which the manufacturer entertains so great a dread as of this, and, therefore, from whatsoever cause it may arise, I would certainly decline to breed from, or keep, any sheep upon which it had once appeared. Two primary causes are assigned for the evil, and supporters of either theory are to be met with amongst practical men. I think, however, we shall have to look to the physiologist or chemist, and the microscope, for a conclusive verdict. The one theory is, that protracted wet weather swells the tips of the fibre, and if this be followed, when the wool is about half or three parts grown, by equally protracted dry and hot weather, whilst at the same time there is a scarcity of grass, the fibre contracts forcibly enough to cause it to rend or split up into its constituent filaments. The other theory affirms that the defect is hereditary, and charges it simply and solely to bad breeding. For my own part, I incline towards the latter theory, as being at least the safer, but I think it probable that atmospheric influences may at all events develope the latent hereditary predisposition.

Locky Wools, as the name denotes, are known by the fleece breaking or parting into separate locks. The top of the wither, behind the shoulder, and at the points, are the parts most subject to this defect, and I have also seen it pervading the greater portion of the ribs. It is not a very serious fault, because so easily detected. When the sheep is shorn, the locks fall separately on to the shearing-floor, and, of course, are put with the refuse wool. Sheep of this kind should be rigidly culled out, as the defect is, beyond doubt, hereditary. Bad breeding originates, subsequent neglect or carelessness in culling will perpetuate it. There is no excuse for this, because at shearing time its existence will be made abundantly manifest, even to the merest tyro in sheep management.

Felty Wool, or wool that has a tendency to felt on the sheep's back, is very similar in outward characteristics to broad-topped wool, and is generally most prevalent after a very wet season, or when the sheep have been starved, and are still in very low condition. Some descriptions of wool are much more liable to "cot," or felt, than others, but I have known sheep felt one year, and be free from the defect in the succeeding year. Quoting from a letter addressed to Mr. Youatt, by Mr. Plint, it appears that "the property of felting is altogether inexplicable, except on the supposition that the extreme surface of the fibre is irregularly feathered, and that by compression these feathered edges become entangled and locked together." (*Vide* "Youatt on Sheep," p. 84.) This is, however, certain, that wool which felts on the sheep's back is generally clothing wool, and as our long combing wools rarely felt, we do not suffer much loss by it ; nor, indeed, have I ever seen sheep that "felt" very numerous in any of the colonies. As, however, the property owes its existence to an organic peculiarity of structure, I should certainly reject any sheep exhibiting a tendency to felting.

Cloudy Wools are evidently similar in their character to felty wools ; adhesion of a like kind, but scarcely so close and intimate, taking place among the fibres, principally at or near the roots, confers upon them the property of felting in a lesser degree. This defect may be observed by holding a staple extended by the two ends, when it will be seen that each fibre is not well defined ; and on examining it with a good lens, the lower part of the staple appears as if it had floss silk mingled with it. If held by the top, and drawn carefully from the point to the root of the staple through a fine comb, more or less of this floss will remain in the comb. In clothing wools, this peculiarity may not be very objectionable, but in combing wools it would be very much so, as it would make an uneven thread, as all the floss would be thrown out with the noils or waste.

Mushey Wool is frequently found on the back or rump of an old, half-starved ewe. It is light, puny, very weak, unsound, and without body. It is also seen on young, but neglected, or ill-managed sheep. It is of the most valueless character, and sheep bearing it cannot too soon be got rid of.

Kemp is a defect very easily detected, and, therefore, more generally understood than other imperfections of wool. It consists of a quantity of coarse white hairs, easily discernible on the face, the forearm, and the inside of the flank; also, in rams, on the scrotum. If, however, kemp confined itself to the parts mentioned, it would be of little consequence, but whenever it is found on the parts described, it invariably pervades the greater portion of the fleece, including the whole of the belly, half way up the ribs, and the lower part of the thighs and forearm. If a small staple of wool be drawn out from any of these parts and held up to the light, at the roots will be seen these short white hairs, from half an inch to one inch in length, or thereabout. These hairs will not take the dye, and consequently the wool containing them can be used only for inferior purposes. The most expert wool-sorters and wool-brokers, many of whom the writer has consulted, declare that kemp lessens the value of a fleece from 40 to 50 per cent. It is my firm persuasion that kemp has been communicated to the Merino sheep by crossing with the Leicester, Cotswold, Southdown, or other variety imported from Great Britain. On some sheep, and especially on the Rambouillet and the German Negretti, there are often seen long, straight hairs about the back of the head, not seldom rising above the level of the true wool. Occasionally this is seen also about the edges of the wrinkles, on the tail, and thigh. This is decidedly kemp, although it does not show itself on the face or other parts usually occupied by the short white kemp. But it is even more objectionable than that which is usually recognised as kemp, because it is equally injurious to the value of the fleece, and there is much more of it on sheep so affected. Yet I have seen, in the course of the last ten years, enormous prices paid, by gentlemen who, I thought, should have known better, for rams with this glaring defect. Repeatedly have I counselled gentlemen to whom I was known not to purchase this kind of sheep, and shown and stated to them the disastrous consequences which must ensue upon putting them to the Australian Merino ewe. "Oh," they would reply, "this is a new breed of sheep entirely. You never saw any of them before, and can therefore know no more about them than we do. They have been bred, at great expense, by the best and most scientific breeders of Europe; and these white hairs that you object to so strongly are—*we are credibly informed*—never transmitted to their progeny." This kind of doctrine appeared to me so novel and surprising, that I was taken quite aback, perceiving that, if it were sound, all my life-long labours had taught me less than nothing; and that I, and all the experienced sheep-owners and managers of my acquaintance had been fostering a pernicious error when we believed that the bad qualities of a sire were more certain to be transmitted to his descendants than the good ones; nay, that the highest of all authorities misled us when it announced the solemn truth that "the sins of the father would be visited upon the children." Neither for the physiological nor for the theological fact shall I attempt to account; of its verity no sane man doubts; and this much I call upon all to avouch who have had, unassisted, to acquire their experience, viz., that it is far easier to produce a flock of bad than of good sheep.

Breachy, Tender, or Wool with a Break in it.—There is, perhaps, no defect which renders wool, and otherwise good wool too, so absolutely useless for manufacturing, and especially for combing purposes, as tenderness or breachiness; and it is my conviction that this defect is more general, and causes greater loss to the country, through the pockets of our sheep-owners, than all other defects in our wools put together. However fine, or however much your wool, in every other desirable quality, may excel, no sooner is it submitted to the wonderfully acute and skilful examination of the European wool-sorter, classer, buyer, or manufacturer, than its deficiency in this respect is detected, and a price is bid for it scarcely exceeding that offered for locks and pieces; in fact, nothing is wanting to reduce your fleece to that class but the solution of continuity, which is sure to take place in the course of the very first manufacturing process to which it is subjected. Except, however, possibly in cases where neglect or mismanagement have been the rule for generations, it is not hereditary; nor is any one breed of sheep more liable to it than another. To these conclusions I have come by repeatedly finding an entire flock affected with break one year and quite free from it the next, in consequence of a change in management. On the whole, it is to be feared that this defect is yearly gaining ground; and I am credibly assured that, for the last two or three years, we have produced more wool of this description than was ever known before. Certain it is that wheresoever this most objectionable tendency manifests itself, sheer carelessness, neglect, ignorance, overstocking, inordinately large paddocks, or scarcity of feed or water, each or all will be found.

When sheep get into very low condition, the pores of the skin contract, and permit only wool of a very fine fibre to extrude. When the feed once more becomes abundant, the pores again expand, and permit the passage of a larger and stronger fibre. In consequence of this, the extremities of the fibres are stronger than their centres, and the wool, upon the slightest strain, snaps at the weakest place—namely, at the portion which grew when the sheep were in the lowest condition. But nothing is so sure to cause a break in wool, or, indeed, in many sheep a perfect stripping or shedding of the entire fleece, as *want of water.* I have often observed, since the paddocking of sheep has become general, that in a paddock of six or seven thousand sheep, from eight hundred to a thousand have entirely stripped the wool off the body. Now, in an extensive paddock, where the sheep are watered at small tanks or wells, small mobs of them are likely to wander to the extreme verges of the enclosure, in search of food (especially in seasons such as the past), and stop there until they are terribly debilitated—half-dead, indeed. This may occur to several small mobs during the hot season, and unless the boundary-rider is not only assiduous and watchful, but expert in readily detecting the absence of portions of his charge, the flock-master will find, on mustering, a large percentage suffering both from *break* and *stripping.* When a few sheep only are thus affected, the evil may arise from some minor cause, such as foot-rot, a stake, bite of a dog (wild or domesticated), and in ewes from a difficulty in parturition occurring in the course either of the true or after-birth, puerperal fever, etc. ; but nothing, I feel assured, is so likely to cause a *total* stripping as want of water. It is well known that fever induces thirst, but in this case, thirst generates fever, and fever infallibly causes, in proportion to its intensity and duration, either partial or total stripping. In stripping there is a complete closing of the pores, whilst in " break" this takes place but partially. This may be easily verified thus :—Confine a sheep or two in a place without food, and keep them there until nearly exhausted. Mark them, and return them to the flock they came from, which we will suppose to be thriving ; and as soon as the marked sheep begin to show signs of improvement, try the wool, and you will find either a break or total stripping. Much will, of course, depend upon the condition and constitution of the animal before you experiment upon him ; and the better the condition, the longer will he bear starvation without injury to the wool. Or, if you prefer a conclusive but somewhat expensive test, put a few sheep into a paddock where there is abundance of feed, but no water, and no shelter from the sun ; keep them there a sufficient time to tell upon their condition, and you will find that a total stripping will ensue. I have, at this moment, in my mind's eye, a certain paddock, so large that the sheep can get as far as 10 or 12 miles from the watering place. Now, in a paddock of such dimensions, with just an ordinary portion of scrub or timber, it is quite easy for the boundary-rider, even if he should observe that some of his charge had not been to water, to miss them when he went to look them up; but it is a thousand to one he would not miss them at all, if (say) 200 or even 500 out of 12,000 or 13,000 sheep failed to come regularly to water. A very grave, and, since overstocking became the rule, a very prevalent error is to have paddocks too large, or to put too many sheep into them. Want of feed, however. will, in the majority of cases, cause only a break in wool, unless, indeed, the sheep are reduced to the lowest stage of poverty, when total stripping will probably occur; but want of water, I reiterate, in hot weather, is certain to make sheep cast their wool. When paddocks are of such inordinate dimensions, and water perhaps only at one spot, the boundary-rider should not permit the sheep to remain on the fences, lest they become attached to such remote feeding grounds. Other reasons there are, too, for this precaution, most of which are well known to experienced managers; but as I do not write entirely, or even principally, for that class, I will just mention risks of disturbance or injury from travellers and their dogs, or of boxing with a neighbour's sheep, should the fence be imperfect. It is mainly from the occurrence of certain wants and deficiencies, upon which he subsequently accumulates experience, that the practical man gathers his most reliable and valuable knowledge; and a fact so strongly corroborative of my theory of stripping came some years ago under my observation, that I venture to cite it in few words. In the north country of South Australia an entire flock got astray, and were not recovered for about a week, during the whole of which time they were without water. Every sheep cast the wool.

One gentleman told me that he, during the preceding year, had two paddocks, each grazing between 4000 and 5000 dry sheep; in one there was abundant feed, but a paucity of water; in the other, scarcely any grass at all, but plenty of water. The sheep in the well-grassed paddock kept their condition throughout the season, but they stripped off their wool by hundreds, and those who escaped stripping had, every one of them, a break in the wool. On the other hand, the sheep that had had plenty of water, although on the very barest of pasture, and extremely poor, retained their wool perfectly, and had scarcely a tender fleece among them. My informant was at a loss to account for this strange apparent anomaly, but after considering the theory here propounded, he candidly admitted that he could call to memory many cases in which no other solution of his difficulty was possible.

Although the defect to which I am about to advert arises from no error in selection or breeding, but rather from a determined perversion of the true principles of those arts, it is one which has done our wools serious injury in the markets of Europe. Within the last few years, people have had a perfect mania for growing wool of extra length; and, in aiming to produce length of staple, they have too often ignored and neglected the equally valuable property of density. The system has produced a great number of faulty sheep, bearing an open fleece, long, perhaps, but thin on the pelt; and this description of wool has invariably a tendency to curl at the tips, which, in the process of manufacturing, causes it to gnarl and break at the tips. A dense wool, of from 2 to 2½ inches long, is far more profitable, because, other things being equal, the dense fleece will be the weightier; it will also protect the bottom, or roots, of the wool from dust. Besides this, wool of 2 or 2½ inches long can be combed just as successfully as longer wool. Therefore, in exercising my professional duties as a Classer, I do not hesitate to say, that I would prefer a staple of two inches long, with density, to one of four inches, which was thin on the pelt. I would also prefer a good heavy fleece of second combing, to one of short and light first clothing, provided the wool grew properly upon it, free from faults, and even, as regards length, all over the body.

Unevenness means any great difference in the length of staple on various parts of the body of the same sheep. For instance, say the wool on the lower part of the shoulder (usually the longest and best portion of the fleece) is 3½ inches in length, whilst that on the ribs has attained only 1½ inches; the difference in length is too great, and you pronounce the sheep "very uneven." Again, compare the shoulder wool with the wool on the thighs; the former is found to be fine, long, and generally of good character; the latter is coarse, harsh, destitute of character, and nearly straight; that is, it has no curve, consequently, it is devoid of elasticity; this, also, is occasionally termed unevenness; so that the word has a somewhat wide application. The wool on the shoulder being usually the best, is taken as a standard of comparison with the other portions of the fleece. Now, the shoulder wool being 3½ inches long, the rib wool should not be less than 2¼ inches. During my long experience I have not met with fifty sheep whose rib wool equalled that of the shoulder, yet it is of the utmost importance that the rib wool should nearly equal that of the shoulder, because the ribs should furnish fully two-thirds of the entire fleece. The top of the shoulder and wither is the part most prone to fall off or shorten ; and I am sorry to say, that having handled a great number of rams during the last few years, I have found that portion of the fleece getting worse every year. I have lately seen rams with three inches of good wool on parts of them, yet only about one inch on the withers ; and the little that did grow there, in small locks, with a top to them like a church-steeple. When wool grows thus short, and in separate locks on the withers, it falls, when shorn, in detached pieces on the shearing-floor ; it leaves the fleece " broken," and is thrown amongst the " locks."

CHAPTER XVII.

TRUENESS, OR EVENNESS, OF FLEECE.

In the first place, then, trueness, or evenness of fleece means that the whole of the various parts of the fleece should have, as nearly as possible, a uniformity of character; that is, as regards fineness, length of staple, density, softness, and, some add, in the structure and condition of the fibre itself. Now, these attributes, when present in an eminent degree, leave little else to be desired, as any sheep possessing them, with the addition of a symmetrical frame, may safely be pronounced a perfect sheep. I must admit, however, that I have never seen a sheep that was incontestably and perfectly true and even; so that when we say that a sheep possesses a fine even fleece, we speak comparatively only. Always presuming that the wool to be inspected is really a fine wool, we first examine the shoulder, at the part where the finest and best wool is usually found. This we take as the standard, and compare it with, in turn, the wool from the ribs, the thigh, the rump, and the hinder parts; and the nearer the wool from these various portions of the animal approaches the standard, the better. First, we scrutinise the fineness, and if the result be satisfactory, we pronounce the fleece, in respect of fineness, very "even." Next we inquire into the length of staple, and if we find that the wool on the ribs, thigh, and back, approximates reasonably in length to that of our standard, we again declare the sheep, as regards length of staple, to be "true and even." We next desire to satisfy ourselves of the density of the fleece, and we do this by closing the hand upon a portion of the rump and of the loin wool—the fleece at these points being usually the thinnest and most faulty—and if this again gives satisfaction, we signify the fact by designating the wool "even" as respects density. Now to summarise these separate examinations:—If you find the fleece of nearly equal fineness from the shoulder to the thigh; of nearly equal length at the shoulder, rib, thigh, and back; and of equal density at the shoulder and across the loins, you may conclude that you have a nearly perfect sheep. But there is still another defect, which is termed unevenness;

and this is when the fibre differs in thickness, or diameter, at its centre and extremities. It is asserted that both extremities of some fibres are no less than 25 times the thickness of their centre. The statement would seem to make a somewhat heavy demand upon our credulity, but the authorities are of high character. Such wool, however, could be produced only on sheep in a very unhealthy state; and as it is not classed among the varieties of unevenness, but rather under the head of " Breachy, or Tender Wools," it will be found considered in that relation.

I have observed that, even in some of the best flocks in Victoria, the wool has a tendency to become thin and loose, especially on the top of the back over the loins. This may be owing to divers causes; first, to breeding from sheep already defective in this respect; and, secondly, from that portion of the animal being more exposed to rain and sun, which deprive the fleece of the protection and nourishment naturally afforded by the yolk.

CHAPTER XVIII.

ON OVERSTOCKING.

THAT the season of 1868-9 was, for stock of all descriptions, one of the worst ever known, will be admitted by the oldest colonist or stockholder in Victoria. The losses of sheep, in particular, have been enormous, and this notably in the northern part of the colony. The squatters themselves, however, are far from blameless, since, had they refrained from overstocking, the drought would have proved comparatively innocuous. The Banks and Loan Companies also have borne no inconsiderable part in fostering the growth of this most pernicious system. Formerly, and not so very long ago, the only question put to the applicant for "an advance" was, "How many sheep have you?" But the Banks and Loan Companies have, happily for the country at large, burned their fingers by some such transactions, and now a second and far more essential query follows, viz., " What sort of sheep are they, and what are the carrying capabilities of the run?" We have, indeed, at the present time, in Victoria, millions of sheep not worth the grass they eat ; that is to say, the fleece will not pay interest upon prime cost of the animal, with rent, assessment, shearing. and other inevitable working expenses. Not only has this fatuous practice of overstocking led to total loss by death, but it has, by sacrificing quality to quantity, laid the foundation for evils of a yet more lamentable and permanent kind—to wit, serious deterioration both of the breed of sheep and of the wool which they bear. The drought of 1868-9 will long be remembered with horror, but let us acknowledge that, even from a dispensation so severe, an element of value may be elicited ; and that, whilst contemplating the sad wreck of property caused primarily, no doubt, by the drought, the squatter will never again, by overstocking, put it out of his power successfully to meet adverse seasons with that judgment and skill which we are graciously permitted to acquire for the purpose. No error has been more common, or attended with more disastrous consequences to stock-owners than overstocking. Even the very persons practising it admit its baneful influence ; yet, perhaps, for reasons into which we need not pry, still persistently cling to it. It is, beyond a doubt, destructive to both sheep and run, and that in a great many ways. For example, people at this very time are complaining and lamenting over the diminished size of our Australian Merino, yet they are taking the most effective measures in their power to render him smaller still, for nothing will so surely accomplish this as starvation. Time was when we could send our fine-woolled Australian Merinos to market weighing on an average 80 pounds, and these were really well-bred sheep, not mongrels, such as are great numbers amongst our existing flocks. Yet, notwithstanding that many Leicesters, Cotswolds, Teeswaters, etc., have been introduced for the purpose of giving our sheep bulk, in place of the Australian Merino becoming larger, he has dwindled down to a mere dwarf. It is now considered a rare thing to see, even in the Melbourne market yards, a flock of sheep averaging 60 pounds in weight. What has occasioned all this? Why, overstocking, and nothing else. There are very few Victorian flocks that have not some of the Leicester, Cotswold, or Lincoln blood in their veins; and these races being so much larger than the Australian sheep, we certainly ought to have had an increase in point of size; whereas it is patent that we have lost in weight of carcase, as well as in quality of wool. This is simply the result of the runs being overstocked, and the sheep consequently starved.

Overstocking, moreover, reduces the quantity of wool which each individual sheep would produce if well fed, and that to the extent of, at least, 30 per cent. I speak advisedly when I make this statement; and I really believe that I am 20 per cent. under the mark, in cases where the animal is reduced to a very low state of condition. Again, we lose very considerably in the quality of our wools; since it is an undoubted fact that starving a sheep will surely be followed by a break or a tenderness of the whole fleece; nay, in many cases, a perfect stripping will ensue. (*Vide* p. 86, on Stripping.) But this is not all the ill done by overstocking, even to the animal and its wool; for, as I have elsewhere shown more in detail, the short grasses and sand wear down and destroy a sheep's teeth before he is in his prime. (*Vide* Chap. V., on Sheep's Teeth.)

The injury done to the run itself by the practice of overstocking is very serious; amongst other reasons, because it both propagates and encourages weeds and inferior grasses. It is well known that there are certain herbs and grasses to which sheep have a great repugnance,—which, indeed, they will not touch, even though they be starving. Now, the animal, in selecting his food, removes from around these useless or noxious

plants every particle of vegetation; and this procedure, just
like the weeding of a garden, leaves the objectionable plant in
sole possession of the land, where it grows and seeds with ten-
fold vigour and fecundity after the removal of its rival, the
nutritious and useful herbage. The valuable grass gets no
chance to seed, the worthless stuff seeds abundantly, and
grows luxuriantly, until, from a patch of a few acres, it gradually extends itself all over the run. On a well-grassed—
that is to say, upon an understocked, or even a fairly-stocked
run, this could not take place, because the vigorous growth of
the useful grasses would choke or, at least, very much check
the increase and spread of the useless ones.

There is, I admit, some difficulty in arriving at the exact
number of sheep that a run will carry, and keep in top condition, in consequence of the varying of the seasons with respect
to rain; but if those who have runs impoverished by overstocking, desire to change their system, let them at once
reduce the number of their stock to a quantity which the run
will carry well in an unfavourable season, and keep them down
to that limit until the run has recovered itself, the grasses
seeded, and the sward thickened. Should you then have more
grass than you require, purchase a flock or two of store
wethers to fatten, and so utilise your surplus feed, without
putting a permanent stress on the run.

The system of sheep management has, of late years, been much
improved; especially, by fencing and subdividing the runs, by
conserving the water in dams, tanks, and wells, by hot water
and spout washing, and last, but not least, by the employment
of skilled persons to direct the shearing, classing, and getting-
up of the wools, whether these latter most important operations
be performed in the squatter's shed or the wool-broker's store.
But the time has now arrived when, if squatting is to be a
profitable business, much greater skill and discrimination than
has heretofore been displayed in selecting and breeding must
be exercised. The squatter has had, and still has, many
depressing influences to contend against; for instance, the low
price of wools and mutton, and no sale whatever for surplus
(or "store") stock; a pitiless mangling of the country by free-
selection under the 12th and 42nd clauses of the Victorian
Land Act; and, to crown all, a season of unparalleled ad-
versity. The old system has palpably failed in contending
with these manifold obstacles, but surely they are not in-
superable. What we want is, in the first place, a better class
of animal, and these can be obtained only by an improved
system of selection, and by vigorously culling and rejecting

from our flocks every objectionable sheep, of either sex, short-woolled, coarse-woolled, badly-shaped, or mongrel-bred. I have rarely seen an overstocked run which did not produce wool with a break with it. Although I attribute to overstocking much of the deterioration which confessedly has befallen our sheep and wool, yet I feel bound to state that there are amongst our squatters eminent exceptions to the general rule; for example, the Messrs. Learmonth, Currie, Shaw, Bell, Cummings, Mackersey, Gray, and some others, whose names I cannot at this moment recall.

The description of sheep most suitable for our climate and pasture is proved to be the Australian or acclimatised Merino; and it is by the careful selection, breeding, and culture of this animal that the gentlemen I venture (without their permission or cognisance) to name have achieved successes, patent, not only to the colonies, but to the wool markets of the world. There are many faults or imperfections in sheep which do not attract the notice of casual or inexpert observers; among which may be named kemp, toppiness, stripy, unevenness, tenderness, felty, and others, which, although pointed out by many writers on colonial sheep-farming, have not hitherto been clearly and exhaustively explained. It must be mortifying to the uninitiated seeking information and guidance from such works, to be told that his wool is heir to all these natural ills, and yet not to be informed how he may verify the fact, and guard against its recurrence. To parties so situated, it has been the writer's aim, in other chapters, to supply the desired instructions.

CHAPTER XIX.

THE EFFECTS OF CLIMATE ON WOOL.

IT has often been stated that the salt-bush countries cannot produce superior wools, because either the excessive heat or the saline herbage, or both combined, render the wool harsh, coarse, and thin on the pelt. The first of these objectionable characteristics I admit, and its cause I have already indicated in my remarks on yolk, and its influence on wool. The two latter, I think, may be traced to other causes; not that I would infer that wool can or will ever be grown in the salt-bush countries equal to some of the wools produced in the Western districts of Victoria, as the salt-bush wools will always lack the softness and lustre peculiar to the Western wools. Coarseness and thinness, however, can be traced to nothing but want of judgment in breeding, and an insane desire to " stock-up," which in most cases means "overstocking." The occupation of "new country" in Queensland, New South Wales, Victoria, and South Australia, has frequently been effected under the observation of the writer, and he has remarked that, although a party may take up for 5,000 or 6,000 sheep sufficient land to feed ten times the number, yet he is never satisfied until he has effected this pernicious " stocking-up." In pursuit of this object, everything is bred from that can rear a lamb or beget one ; and here is the grave error, and the direct cause of coarseness and thinness in wool. " Here," say they, " is abundance of grass and water ; all we want is sheep to stock it." A few cheap lots, consisting of coarse or inferior sheep, are bought, to effect this object, and on the principle that it is better to have bad sheep than none at all, they go on breeding indiscriminately for numbers only. In a few years, the squatter finds that the original stock is exhausted, and that he has indeed " stocked-up" with a lot of sheep not worth the grass they eat. Next, they perceive, perhaps from a study of their agent's balance-sheet, that the wool is coarse, harsh, light, and full of faults, and also that the price obtained for the " clip" is anything but satisfactory. Forthwith they complain of " the climate," denounce the broker who sold the wool, and bespatter with censure all, except themselves, who had any connection with the run, the sheep, or their fleeces ; but it never for a moment occurs to them that their own cupidity, or their ignorance of the principles of breeding, has been at the bottom of the mischief. And this is why we hear people say, that good wool cannot be produced on salt-bush country. Now, I have not the slightest hesitation in stating that salt-bush runs can, and in some instances do, produce as heavy and as valuable fleeces as do any other portions of the Australian colonies. I mean in point of money value, taking fleece for fleece. On the Cobran and Thule stations of Messrs. Wolseley, Gibbs, and Co., in Riverina, I saw, in 1868, some rams shorn, whose fleeces weighed from 9½lb. to 14¼lb. (in the grease, of course), and these fleeces were of really first-class quality and character. A fact such as this occurring on stations so large as the Cobran and Thule, I hold to be conclusive. The salt-bush squatter has, indeed, some advantages which he of the grass country does not possess, or, at least, not in a similar degree. Sheep fed on salt-bush country are always healthy, not subject to fluke or foot-rot; their wool is usually sound, although perhaps a little harsh, in consequence of the deficiency of yolk. The ewes are generally very prolific, and produce a larger number of lambs; and surplus stock can nearly always be fattened. If, then, the squatter has any ewes that are not up to the mark as regards wool, he can fatten them, and send them to market. There is really nothing whatever to prevent the salt-bush squatter producing sheep with very heavy fleeces of good sound wool, the money value of which taken together would be fully equal to that of the general run of sheep and fleeces in even the western districts of Victoria. On the other hand, the squatters of the back blocks labour under certain disadvantages, the principal of which perhaps consists in the fact that large bodies of sheep have to water at one well, and are consequently enveloped in a cloud of dust going to and returning from water. Good wool cannot be produced under such adverse circumstances; as indeed nothing tends more to destroy both wool and sheep than dust; the heat causing the yolk to run, and the fine dust absorbing it, rendering it dry and harsh, whilst the exudation from the pores of the skin is impeded, and the health of the animal suffers in much the same way that the health of the human being is affected by the stoppage of those important safety valves. The want of shade is another cause of deterioration, both of sheep and wool. I have seen large paddocks, consisting entirely of plain, and not a single tree to protect the sheep from the influence of the hot mid-day sun, the thermometer standing at 130 degrees. What wool or sheep could stand this without grave injury? Sheep absolutely require some protection from the heat of the sun, to which they might retire and quietly camp during the heat of the day; and rising refreshed as the day waned, feed vigorously and with health-conferring zest during the night. Moreover, sheep that have plenty of good shelter or shade will not require near so much water; without shade they will walk all the flesh off their bones, and destroy their wool by the dust they raise in ceaselessly marching along their own well-worn tracks, almost in Indian file. The half-watered and scantily sheltered back-blocks can never be continuously used with advantage; they are adapted merely for sheep to occupy during the winter months, when there is little dust, and surface water and feed easy of access. With these exceptions, the truth is that salt-bush squatters have little cause of complaint, except such as arises from causes which I need not a second time specify. Nature has done much for them, and would assist them in every respect would they but assist themselves. Never could the adage have a more apposite illustration, "Help yourself, and heaven will help you." Proofs are not wanting that the salt-bush country can produce good and valuable wools.

In the year 1845 I had occasion to visit the Darling Downs, when I examined a large number of sheep, and amongst them some of the best then to be met with in that fine pastoral tract, to wit, those of the Messrs. Leslie of Canning Downs, the Gore's, Forbes Bros., Hughes and Isaacs, the Gammies of Talgai, and many others, all of whom possessed sheep as faulty as sheep could be in respect of wool. It was at that period the received dictum that the climate was too hot for the growth of good wool. Mr. F. Brachar, however, who was the chief superintendent of the Clyde Company, thought otherwise, and proceeded vigorously and perseveringly to demonstrate the correctness of his views. That gentleman adopted at once a careful and judicious system of selection, etc., and when, some eight or nine years subsequently, I again went through the same flocks, I found sheep that I had rarely seen equalled for weight of fleece and quality of wool. The Leslies, the Gammies, Mr. Duchar of Glen Gallan, Messrs. Kent and Weinhalt of Johndarian, purchased rams of Mr. Brachar, adopted his system, and followed his advice; and, finally, the Darling Downs at this day produces as good wool as is grown in any of the Australian colonies; the Talgai wool, indeed, realised, in 1867, from 2s., 2s. 6d., to 2s. 10d. per lb. in London; certainly not bad for salt-bush, or, as it is termed, "hot-country" wool. One word of counsel I will, with all respect, tender to the salt-bush squatter. Do not aim at breeding large sheep, keep them short on the pins, and their bodies close to the ground. The moment you commence to breed for size you go wrong; and never, under any circumstances, breed from long-legged ewes or rams. The warm climate itself has a tendency to produce "legginess," but you should be satisfied if your six and eight-tooth wethers will weigh from 60 to 70lbs. This is quite enough for a fine-woolled sheep to weigh, fed upon the natural grasses.

Overstocking in paddocks deteriorates sheep and wool, especially the latter, in a way very similar to that which occurs round wells in the back country; that is, the sheep keep the ground so bare that they are almost constantly enveloped in clouds of dust. Sheep bred in a salt-bush country should rarely be sent to a grass country after they are two years old, as they seldom thrive when so removed; they require acclimatising. But sheep sent from a grass country to salt-bush will fatten at almost any advanced age, even without a single tooth. The grass country squatters have many obstacles to contend against that the salt-bush men know nothing about. It is true that they can produce wool which perhaps no country in the world can compete with, but it is only in favourable seasons that they can fatten their surplus stock. Their sheep are not so long-lived. Their increase is not to be compared with the increase on the salt-bush stations; in fact, the decrease is greater and the increase less. They have fluke, as well as that abominable and expensive disease, foot-rot, to struggle with. They have nothing to rely upon but the superiority of their wool. And, finally, sheep in grass countries are more liable to disease of all kinds than sheep fed on salt-bush. It consequently requires more care and judicious management of the most costly kind to make the station pay.

CHAPTER XX.

WOOL PRESSING.

THIS operation is so simple that very little need be said about it. I would, however, mention that I consider it a great mistake to press your wool too tightly; in other words, to put too many fleeces into a bale. It is a too general practice to aim at making what is called a neat bale, and this is effected by forcing into about two-thirds of the bale a quantity of wool amply sufficient to fill the package. About 230 or 240 lbs. of washed wool is as much as should be put into a bale, and this should occupy the whole length of it. By pressing too tightly, the fleeces get jammed into one another, almost as if they were felted together, and not only is there great difficulty in disentangling and separating them, but it is next to impossible to do so without more or less detriment to the wool. The man who immediately superintends the press is the one you must watch, for these persons generally take a pride in turning out these very pernicious "neat bales."

Not in one shed of a hundred are the fleeces put into the bale with sufficient care, and on the best system. You will see them pitched, not even singly, but two or three at a time, up to the man in the press, who sometimes catches them, and often fails to do so, when down they come on to the dirty floor, and perhaps fall open; they are then hastily and improperly re-rolled, and in this state squeezed into a felty mass in the pack, plus sheep's dung, bits of stick, and any other filth they may pick up. Those fleeces, when the manufacturer sees them, will scarcely elevate the reputation of your brand in the market. Let the fleeces be handed, one by one, to the presser, and put carefully, and with some attention to regularity, into the pack, say four or five on each side, according to the size of the fleeces, and on no account permit any to be unrolled or opened. It is a common thing to see wool broken up, and torn almost into pieces, in the process of packing, especially when topping-up or putting the few last fleeces into a bale. No more wool should be placed in the pack than will reach the top edge of the press, or than the monkey can take down without causing it to bulge out.

The manufacturer has an objection to fleeces being tied with string, as portions left in the wool injure the delicate machinery. It is not at all necessary to tie your fleeces, provided they be rolled, handled, and placed in the pack as I have already advised.

MELBOURNE :
CLARSON, MASSINA, AND CO.,
GENERAL PRINTERS
LITTLE COLLINS STREET EAST.

WOOL.

GOLDSBROUGH'S WOOL WAREHOUSES,

Bourke and William Streets, Melbourne.

R. GOLDSBROUGH AND CO.

Beg to inform the Woolgrowers and Merchants of Victoria, New South Wales, South Australia, New Zealand, &c., &c., that the undermentioned firms act as our Agents, and attend to the forwarding of all consignments entrusted to our care, pay carriage, freight, and insurance (when advised) :—

Sandhurst—
Messrs. R. GOLDSBROUGH & Co.

Ballarat—
Mr. E. J. STRICKLAND, Lydiard-street

Echuca—
Mr. F. PAYNE.

Geelong—
Mr. T. J. PARKER.

Hamilton—
Messrs. CROAKER, SCOTT, & Co.

Castlemaine—
Messrs. W. M'CULLOCH & Co.

St. Arnaud—
Mr. H. GORMLEY.

Port Albert—
Mr. WM. HOWDEN.

Belfast—
Messrs. GRANT & HARPER.

Warrnambool—
Mr. R. B. PATERSON.

Portland—
Messrs. CROAKER, SCOTT, & Co.

Wahgunyah—
Mr. ROBERT LOWES.

Deniliquin—
Messrs. MORT & WATSON.

Hay—
Messrs. FORSYTH & LOUGHNAN.

Wagga Wagga—
Mr. GEORGE U. ELLIOTT.

Wentworth—
Mr. WM. GUNN.

Port McDonnell, S.A.—
Messrs. THOMAS MUST & Co.
   „   N. P. LORD & Co.

Penola—
Mr. FRED. LAW.

Robe Town, Guichen Bay—
Messrs. ORMEROD & Co.

Mount Gambier, S.A.—
Messrs. N. A. LORD & Co.

Adelaide, S.A.—
Mr. JOSEPH DARWENT.

Invercargill, N.Z.—
Messrs. M'PHERSON & Co.

CHARGES:

Receiving into **Store, Warehousing, and Storage** (one-eighth of a penny per pound)

Repacking sample bales (if any) 5s. each.

Insurance whilst in Store, one-eighth, or 2s. 6d. per cent. on market value. (See note at foot.)*

Commission on Sales effected, one and a-half per cent. ... ...   1½ per cent.
Under £200   ... ...   2½   „

* INSURANCE.—This charge is made in all cases, unless the owner specially instructs us not to insure. Wools are covered under our open policies from the moment they enter our warehouses.

R. GOLDSBROUGH & CO.,

WOOL BROKERS, AND STOCK AND STATION AGENTS.

ADVERTISEMENTS.

TO THE
SQUATTERS AND STOCK-OWNERS
OF
VICTORIA, NEW SOUTH WALES, & QUEENSLAND.

MELBOURNE, 1ST JANUARY, 1870.

GENTLEMEN,

Believing that, from the largely-increased quantity of live-stock passing through the Melbourne Market, there is ample room for another firm of Salesmen, we have

COMMENCED BUSINESS
AS
SHEEP & CATTLE SALESMEN,
AND
STATION AGENTS,
Under the style of
PECK, HUDSON, & RAYNOR,

and trust that our very long experience in the business will secure to us a fair share of support. Our MR. PECK has been Cattle Salesman to MESSRS. DAL. CAMPBELL AND Co., for eight years; our MR. HUDSON, Sheep Salesman for the same firm, and for MESSRS. ETTERSHANK, EGGLESTONE, AND Co., for eleven years; and our MR. RAYNOR has been upwards of sixteen years in the employment of KAYE AND BUTCHART and their Successors.

All Stock consigned to us for Sale will have the personal attention of one of the firm, and every effort will be made on our part to keep down to the lowest possible amount the charges and expenses, and in every way to give satisfaction to those who may favour us with their patronage.

Account Sales will be rendered and proceeds paid for all Fat Stock on the morning after the day of sale.

We are, Gentlemen,
Your obedient servants,

J. M. PECK.
WILLIAM HUDSON.
T. R. RAYNOR.

ADDRESS—
47, BOURKE STREET WEST,
MELBOURNE.

A TREATISE

ON THE

AUSTRALIAN MERINO.

BY

JOHN RYRIE GRAHAM.

MELBOURNE:

CLARSON, MASSINA, AND CO., AND ALL BOOKSELLERS.

OPINIONS OF THE PRESS.

THE publication of the first treatise on colonial sheep, at least the first in any way worthy of the title, is an event which deserves more than a passing notice. On no subject has a book containing sound information and advice been so frequently asked for, and now, one worthy of commendation has at length appeared. —*Australasian*, July 16th, 1870.

A valuable and practical work. To the farmer who must soon compete with the squatter, the work ought to prove very valuable. It tells the story of a life's experience.—*Ballarat Star*, July 8th, 1870.

Mr. John Ryrie Graham, a gentleman long associated with sheep-farming in the colonies, has just published a most useful treatise on the Australian Merino. All will hail with pleasure a work which must have for its effect the dissemination of much useful knowledge. It is written throughout in a plain, unassuming, common-sense style, which stamps the writer a thorough expert in theory and practice.— *Bendigo Advertiser*, July 7th, 1870.

No part of the treatise is better deserving of unprejudiced and close investigation than the chapter on "Scab and its Cure." We shall have soon a numerous class of sheep-owners among small freeholders and selectors. To these persons the book will constitute a reference of great value. Our sheep farmers, great and small, need not distrust their guide.—*Economist*, July 22nd, 1870.

Throughout the work, the reader is led to feel that the author is repeating his every-day experiences, and we commend it to the perusal of all who own or have any intention to own a flock, or a score of sheep.—*Leader*, July 28th, 1870.

A most important and valuable work, which Mr. Graham's long experience in sheep-farming has enabled him to deal very exhaustively with.—*Record*, July 21st.